Matthew Boulton College Library

Engineering Data Management

A Guide to Successful Implementation

Kenneth G. McIntosh

McGRAW-HILL Book Company

London · New York · St Louis · San Francisco · Auckland · Bogotá
Caracas · Lisbon · Madrid · Mexico · Milan · Montreal · New Delhi
Panama · Paris · San Juan · São Paulo · Singapore · Sydney · Tokyo
Toronto

Published by
McGRAW-HILL Book Company Europe
Shoppenhangers Road · Maidenhead · Berkshire · SL6 2QL · England
Tel 01628 23432; Fax 01628 770224

British Library Cataloguing in Publication Data
McIntosh, Kenneth G.
   Engineering Data Management: A Guide to Successful Implementation
   I. Title
   670.285

   ISBN 0-07-707621-4

Library of Congress Cataloging-in-Publication Data
McIntosh, Kenneth G.
   Engineering data management: a guide to successful implementation
   /Kenneth G. McIntosh.
       p.   cm.
   Includes bibliographical references and index.
   ISBN 0-07-707621-4
   1. Computer integrated manufacturing systems.   2. Database management.
   3. Production management—Data processing.
   I. Title.
   TS155.63.M37   1995
   670'.285—dc20   94-31154
                                                              CIP

Typeset by Create Publishing Services Ltd, Bath, Avon

12345 BL 98765

Printed and bound in Great Britain by Biddles Ltd, Guildford, Surrey

Contents

Preface

From the beginning, life has had many variables. Each person's existence has so many internal and external variables acting on it that a unique journey through life results. This makes it impossible to predict exactly what will happen. There are, however, two aspects to life that can be guaranteed:

- Things will change.
- Each of us will die.

We cannot do much about the second, but we can learn to accept the first. We can accept, tolerate, mould, predict, utilize, plan and accelerate or decelerate its pace. Change affects all aspects of our lives—our expectations are heightened and our working lives transformed. One of the main reasons for this today is technology.

The advances made in all aspects of technology have accelerated at an incredible pace over the past few years and show no signs of tapering-off. In line with the technological 'revolution' there have been other aspects which have conspired to alter the face of manufacturing technology and set a blueprint for the future. These include major advances in computing technology, a changing world marketplace, fluctuating interest rates and more exacting customer demands, recession and a dramatic shift in the world's centres of manufacturing. This change in the world's centres of manufacturing, coupled with the increased competition resulting from global trading pressures, have been of major concern over the past few years. This has resulted in many traditional industrialized nations formulating long-term strategies to move their industrial and manufacturing base from decline to growth.

Many manufacturing and service industries are now aware of this need and are being forced to consider changing long-established cultures and methods of operating, simply in order to remain in business, never mind grow significantly. In fact, it could be said that the only certainty in business today is change—permanency and constancy are rarely found in the modern marketplace. The correct application of the need to become more responsive to change and embrace new philosophies, techniques and methodologies has resulted in an increasing number of descriptive terms such as 'world-class performance' and 'business excellence'. These in turn focus on what are becoming common business goals and objectives:

- The need to increase quality
- The need for increased flexibility

- The need to control costs
- The need to reduce time to market

A wide range of management techniques and philosophies, supported by a convergence of computer technology and applications, are being used to address these issues. Solutions such as CAD, CIM, MRPII, AMT, JIT, TQM, etc., are now well known and established, and are being continuously refined, extended and increased in number.

One of the newer solutions to join this more established collection is Engineering Data Management (EDM). The adoption of more formalized techniques relating to the collection and management of engineering data has been growing rapidly over the past few years and has now matured into an identifiable solution in its own right. It can now claim to be one of the key tools which, when used as part of an overall integrated manufacturing strategy, will have a significant effect on the ability of a company to control and benefit from its internal engineering database.

This control, when used correctly, can manifest itself in related business benefits such as quicker response to market pressures and improvements in the quality of products and related services.

A common mistake in adopting techniques such as these is that they are often implemented from an isolated and/or technical viewpoint. What must be remembered is that they are only tools to enable these business benefits to be achieved and concentrating on them in isolation will only automate existing practices without improving the overall view.

I decided to write this book because I felt that there was a lack of practical books on Engineering Data Management technology. I am a qualified engineer although my work now primarily focuses on information technology related to engineering. Most of the books I have read on information technology and computer-related engineering solutions tend to concentrate on the technical aspects and are usually presented in a theoretical manner. I wanted to write a book on Engineering Data Management from the 'real-world' viewpoint of engineers and managers within a company, where internal policies and procedures, culture, politics and the manufacturing environment concerned are as important as the technology itself.

This book therefore seeks to explain both the technical and business-related aspects of EDM to allow a clearer understanding to be made regarding the subject. It is presented, as far as possible, in a non-technical manner, thus easing the introduction to the subject, and only a basic awareness in engineering and computer applications is assumed. In this way it is hoped that it can also be used by other manufacturing industries to act as a key tool to aid in their drive towards a more efficient, profitable and successful enterprise. Only by the correct adoption of techniques such as EDM can we hope to increase our manufacturing base, reverse the decline in manufacturing industry and enable companies to achieve business excellence and world-class performance.

The material used to generate this text has been gathered over many years, in many different environments and from many different sources. I do hope that this diversity helps in giving a wide perspective to Engineering Data Management and will help those new to the subject by providing sound and practical guidance.

This book could not have been written without the expertise and practical knowledge which I have gained in working with a large number of fellow engineers and technology specialists. I would like to thank all those colleagues, past and present, whom I have worked with and who have contributed, in whatever way, to the writing of this book. Thanks must also go to my editor at the McGraw-Hill Book Company and to those who gave their time and knowledge to proofread the initial text. My deepest thanks must also go to my wife Ann for her unwavering commitment, support and encouragement throughout the entire period of the project, and to Scott and Ross who often did not see me for what they consider to be very long periods of time. Finally, my very special thanks must also go to my parents for their continued support and for giving me the chance to undertake the journey in the first place.

Acknowledgements

The author gratefully acknowledges the help received in many forms and from many sources in the generation of this book. Numerous friends and colleagues have contributed in some way—some knowingly and others not. I would like to thank them all. Special thanks should be extended to those who gave their time to proofread the original text: Ian French, Len Northfield, Bill McIntosh MBE, and Ian Emerson.

Help was also received in the form of contributed material, information, corrections and improvements, samples and permission to reproduce diagrams and figures, from the following:

Keith Nichols, EDS-Scicon
Ed Miller, CIMdata Inc.
Barry Brooks, PA Consulting Group

Andersen Consulting
Computervision Ltd
Cranfield School of Management
Frost and Sullivan
Institute of Configuration Management
University of Manchester Institute of
    Science and Technology
Wicks and Wilson Ltd
ServiceTec Infographics
Dr Geoffrey Boothroyd
KPMG Management Consulting

Hewlett Packard
Touche Ross Management
    Consultants
Manufacturing Systems Portfolio
    Limited
Oliver Wight UK Ltd
Ian Hugo and Data General Ltd
Price Waterhouse Management
    Consultants
Computing magazine
Pafec Ltd
Sherpa Corporation
Bull Information Systems
Eigner & Partner GmbH

1

Introduction

1.1 EDM—what is it?

We are all aware that the world we live in today is vastly different from the one which existed only a relatively short time ago. Many aspects have changed, and it is not just the change itself that is important but also the speed at which it must be managed to stay on top of the market. One of the main areas of change which we have all witnessed is that of an unprecedented increase in technology. The 'technological revolution' is rather an ageing phrase but it is still relevant. For example we have video recorders with multi-function timers, high-fidelity stereo systems with comprehensive remote controls and sound emulators, microprocessor controlled household goods of every description.

One of the results of this increase in functionality and performance is that the data required to develop and specify products today has increased at an even faster rate. This would probably be acceptable if this was the only factor in the equation, but unfortunately this is not so. Many other factors have conspired to produce even more data. These include the increasing use of standards (each product must detail those it conforms to and any exceptions), product liability laws (specifying exactly what the product will and will not do), the increasing awareness of quality issues (producing quality user manuals), documentation and after-sales service, etc. The number of documents that require to be managed can be immense. For example, a typical automobile has more than 100 000 parts each with several related documents such as detailed design files, NC programs, simulation results, etc.—a fact made even more frightening when one considers that various studies have confirmed that approximately 25 per cent of all documentation being referenced at any one time is out of date.

This increase in data, which can exist in many forms—for example 'documents' such as papers, drawings, charts, etc., and computer files, disks and cartridges—is being experienced in all manufacturing and service sectors but is especially apparent in defence and defence-related industries. This is adequately demonstrated by the US Navy warship example, where the associated design and service-related documentation actually consumed more space than

1

the warship itself. Aerospace industries have also quoted figures equal to or greater than 1:1 in relation to the weight of associated documentation versus the weight of the product. This, incidentally, is one of the major reasons for the emergence of the Continuous Acquisition and Life-cycle Support (CALS; previously known as Computer-aided Acquisition and Logistic Support) initiative by the US DoD. CALS is aimed at providing standards and controlling the generation, storage and transfer of such information. This initiative is discussed in more detail in Section 5.7.

So, we can see that, in the past, an inefficient documentation, change and control system while undesirable, would not have the same impact as it does today. Hence the increasing interest being shown in EDM, which enables a proper mechanism to exist for the collection, control, dissemination and archival of all data relating to a product or service. Of course, in line with the generally increasing use of computers in all walks of life, computer-related product data has also correspondingly increased—examples such as computer-aided design and draughting (CAD), computational fluid dynamics (CFD), and computer-aided testing (CAT) spring to mind.

A number of initiatives have been made over the years to control and contain both data types, such as traditional database structures, drawing files and databases, Drawing Office Management Systems (DOMS) and manufacturing-related data and planning systems. While providing localized productivity benefits to the appropriate selected area, these approaches did not take a unified, holistic approach to the overall problem and as such the need for a comprehensive system to control engineering data was required.

While the interest and adoption of EDM is increasing, understanding has been held back by conflicting acronyms and names to describe it. EDM is also known as Product Data Management (PDM), Product Information Management (PIM), Technical Document Management (TDM), Technical Information Management (TIM), Engineering Management System (EMS), CAD/CAM Database Management and others. It also has many derivatives or subsets which can also confuse the issue, such as Drawing Office Management Systems (DOMS), Engineering Records Management (ERM), and even Engineering Document Management which is an EDM subset that relates to the control and management of engineering documents. This, together with the lack of a clear definition of what EDM actually is, is slowing down its acceptance and use—it can mean many different things to different people and any definition is likely to provoke discussion. In its widest sense it can be described as 'the systematic planning, management and control of all the engineering data required to adequately document a product from its inception, development, test and manufacture, through to its ultimate demise'. This process is shown diagrammatically in Fig. 1.1.

In summary, we are talking about 'engineering data' control to completely describe and support a product through its development, manufacture and

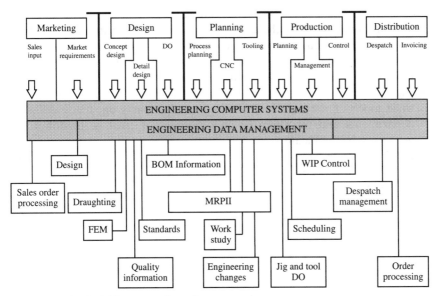

Fig. 1.1   EDM within a manufacturing enterprise

operational life. If you are involved in the production of any type of product or service, from the smallest or simplest product to the largest product or project, you will have a need for some form of EDM system—whether it be a simple card index system or a multimillion pound computer installation.

We can see from Fig. 1.1 that there are many data items which, although required to be considered in the manufacture of a product, are not what could be classed as 'engineering-related'. These include items such as sales orders, invoices, despatch and shipping data. Such data items are usually held in other forms of database such as a manufacturing system or an accounting package. EDM therefore has to coexist with other systems and indeed may integrate with them to pass data to/from them.

The need for integration is discussed throughout this book. The days of 'black-box' insular computing techniques are long gone, to be replaced today with 'open', shared data techniques which ensure that re-keyed data and subsequent transcription errors are kept to an absolute minimum.

Although EDM is known by many different names a number of key elements are common:

- *An underlying database*   A modern, structured EDM solution requires some form of database management system (DBMS), which is a computer tool used to store the data concerned and provide a mechanism for manipulating and controlling this data.

- *A data structure*   The DBMS structure (schema) employed with an EDM system should also have the ability to maintain product structure relationships and historical data (i.e. issue history) for engineering change and traceability purposes. The relationships are usually held via a product structure, or bill of materials (BOM) file.
- *Access control*   Various access controls are required on an EDM system since much of the data initiates from, and impacts on, many different departments and therefore the consequences of error or unauthorized access can be significant.
- *Document status flags*   In order to be used successfully, EDM systems have to replicate the way engineers and the manufacturing industry operate—the computer 'mystique' must be hidden. Engineering 'documents' (i.e. packaged selections of data) usually undergo a series of status changes as they proceed through various company departments, for example create/acquire, revise, index/catalogue, store, protect, distribute, track, dispose of. Tools to enable these status changes have to be provided in addition to the above access controls.
- *Data searches*   Many figures have been issued to illustrate how much time engineers spend in routine tasks such as information retrieval—these range from 15 per cent to 40 per cent. Whatever the real figure is, it involves tying-up a trained engineer in a time-consuming clerical task. One of the main functions of an EDM system is to provide the data-search functionality required for today's engineering environments.
- *Change management*   The management of engineering change and making sure that the latest version of modifications, drawings and specifications are in the right place at the right time is one of the most critical tasks facing engineering management today. This process frequently determines as much as 70 per cent to 80 per cent of a product's final costs. An EDM system should be capable of managing the change control process.
- *Configuration control*   An EDM system must be able to provide full traceability over the entire product structure to enable information relating to the serial number or lot/batch of a product or individual component to be retrieved at any time.
- *Project management*   The usefulness of an EDM system can be enhanced considerably if the data 'documents' are directly linked to projects, i.e. individual 'work packages' related to a particular task, project, product or customer, which require to be managed.
- *Message handling*   Most EDM systems provide the ability to flag actions, warnings and messages via their own, or the host computer's electronic mail (e-mail) system. These messages can be used to determine and track the system's current status.
- *File handling, storage and archival*   EDM systems, by their very nature, store and accumulate large volumes of data from a number of sources. They

therefore have to provide the ability to capture and store data, avoid data corruption and provide the ability to easily archive and restore sets of data.

Many of the individual EDM solution suppliers listed in Appendix A will have individual features added to their products to provide unique selling points, but the above functions constitute 'core' EDM functionality of all good-quality current offerings.

1.2 The history of EDM

An excellent example of the need to control information and embrace new technologies which lead to business excellence was discussed during a recent Manufacturing Resource Planning (MRPII) seminar. It centred around that well-known example—the car. Let us suppose we have decided to enter the car market and have spent some considerable time and money in developing our new car. We have one model in our range—a saloon car with a sporty image, available in one colour: black. Our super-keen salesperson is eagerly awaiting the first customer in our expensive showroom with the car prominently displayed. Our first prospect enters and after considerable discussion is clearly interested in buying. Final negotiations on price take place—our first sale is almost made! 'Pity about the wheel trims though,' says our prospective customer. 'I much prefer aluminium alloy wheels—maybe I should buy the Ford down the road as I originally thought.' Help! What is our salesperson going to do? Let the sale go or negotiate round it? If the objection can be overcome then in general the salesperson will offer the car with the alloy wheels to get the sale. Otherwise he or she may lose their job or soon be out of business.

It is a well known fact that a salesperson's main aim in conceding to a customer is to irritate engineering, and this is exactly what happens! Our car was costed on a basic design: the act of changing it to sell the car has not been considered in an overall sense. Perhaps the alloy wheels cost 100 per cent more than the traditional steel ones, but we have already agreed a price to our customer, so we bear the cost as the manufacturer of the product. We may have to re-engineer the drive shafts or wheel hubs to cater for the new wheels, or perform some additional promotional activities to recover our costs by selling more. A new data record will require to be created and a new BOM structure generated, both of which will require to be maintained and managed over the coming years.

So should the salesperson have offered the option or not? Well, product variation is an important and increasing element in today's marketplace. Many companies can offer a greater range of options to provide what the individual customer wants, while retaining an element of standardization, thus helping to contain these 'on-costs' we have outlined. The further decision could be taken to provide exactly what each customer requires, i.e. 'jobbing', which can work well in specific industries with smaller low-volume, highly configurable products, for example baking machines or overhead cranes.

While the decision to provide this option of car may not have been wrong in this instance, it should be taken in a wider context and the decision should not necessarily be made by one person who is not fully aware of the consequences.

Given that a 'configurable' product line is more common in today's marketplace than, say, 10–20 years ago, it therefore follows that the level of associated product data will have also increased. So where, perhaps, a paper-based system would have sufficed in the past for the control of manual calculations, drawings, specifications, etc., the explosion of data required for these options, which is now often held in digital form, means that such systems cannot now cope. This is especially true when we consider that not only do each one of these documents require to be produced, but that they must also be checked, distributed and subsequently managed.

Other factors which have contributed to the increased modularity of designs include:

- Increased design complexity
- Competition forces a faster 'time to market' cycle
- 'Product rationalization' and 'design for production' exercises

One early tool to assist in the storage and retrieval of product data was the use of microfiche and microfilm techniques. These were commonly used to overcome the problems of storing and handling large drawings and drawing sets, but did not address revision control, configuration management or user access. This is because they were duplicating the functions of existing systems and, while they were much easier to operate and interact with, they did not have the luxury of sophisticated computer functionality. Many of these systems were introduced in the mid-1960s and had to rely on punched card data formats for storing part and location numbers. An example drawing aperture card and scanner is shown in Fig. 1.2. These types of tools are still widely used today, albeit with better quality graphics and reproduction together with improved management and control software.

During the 1970s increasing use was being made of computer technology and this was generally aimed at individual 'point' or localized solutions. For example, the Commodore 'PET' personal computer of the early 1970s was sometimes used to provide local drawing filing and information storage databases linked to a drawing microfiche system.

The trend to use more localized automation products such as CAD, CAE and CNC meant that:

- More data was being held in digital form.
- The need for integration became more apparent—re-keyed data can cause transcription errors.

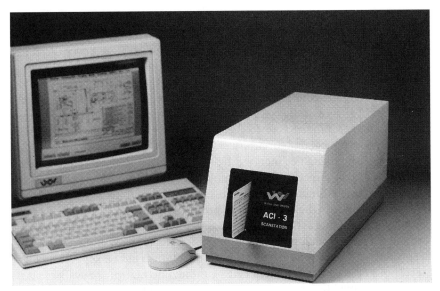

Fig. 1.2 Example drawing aperture card and scanner (*courtesy of Wicks and Wilson Ltd*)

- The 'age of information' brought market pressures to bear—the need to be seen to actively use computers became apparent.

A number of initiatives were born to address these needs. These were mainly related to CAD/CAM vendors but the image and document processing suppliers also enhanced their systems to front-end their hardware. For example, the British CAD/CAM supplier Ferranti Infographics Ltd (now ServiceTec Infographics Ltd) introduced a new breed of product in 1981 which, among other notable innovations, included a module known as ERMS—Engineering Records Management System. This used a networked database model and was seamlessly integrated with the other modules of the system (which was known as CAM-X) to administer and control all the computer system files generated as a consequence of using the modules. At the time this was a leading-edge concept—most CAD/CAM systems before this time were based on 16 bit computers whose outputs were filed and administered using the host computer's operating system only. As such the concept, which everyone agreed was required, was slow to catch on in a practical sense—everyone was busy enough trying to justify spending up to £250 000 on a small number of CAD seats, never mind an intangible database system. But catch on it did, with other CAD vendors such as Computervision offering PDM and Intergraph with their EDMS product, all aiming to satisfy a growing marketplace.

The trends which laid the foundation to these tools have continued and other influences have also helped to raise the general profile of EDM. For example, there are strong indications that the 1990s will be seen as the decade for 'customer service'. In many instances, products and their functionality are converging: reliance on technology alone is not enough to sell over the competition. 'Quality of service' is becoming a main factor in successful selling and this includes after-sales or customer service. In order to provide a comprehensive and efficient service here, configuration management is required: for example, to successfully trace sub-assembly or part revisions, lots or serial numbers for spares requests or to check warranty information or renewal on certain older product configurations.

The increasing penalties incurred by more stringent liability laws, legislation and national directives (for example health and safety-related aspects) when products malfunction also points to a need for configuration management. The increase in the complexity of products has made some of them more prone to failure, while at the same time there is an increased expectation of performance. Perhaps such product malfunctions create the need for a recall of the product. There was a time when such an action could be kept reasonably quiet, but not in today's environment. Although no figures are available it is generally accepted that recalls are far more prominent than, say, 10 years ago, and can prove very costly and damaging. The recent example of a well-known white goods manufacturer can testify to this. A particular model of washing machine, when programmed in a specific sequence, would burst into flames. The embarrassed managing director, when interviewed on television, admitted he had no records to show exactly which models contained the faulty components or who had purchased these models. So an already expensive operation was made even more so by his engineers having to check every single model of that type, from the factory floor, through main stores and distribution warehouses and out to the owners themselves. An EDM solution with a correctly administered configuration management database linked to a customer service system could have enabled a more positive message of 'Only serial numbers .... of machine X contain the faulty components' instead of the more damaging message actually portrayed. Product recall situations such as this where the cost of a modification is at the final production stage or out in the field, for example, can cost half a million pounds or more compared with, say, one thousand pounds if discovered at the initial design stage.

The aspect of increasing costs in making a product also requires to be more closely managed. Whereas in the past it was considered acceptable to design the product and trim excess costs later by manufacturing processes or product re-designs, today's environments require a 'right first time' approach.

Studies have shown that this approach, under the banner of concurrent or simultaneous engineering, can significantly reduce the costs of changes associated with a particular design. Because the majority of costs are actually committed in the early stages of a product's life-cycle, getting the initial design correct at the outset reduces the number, and therefore cost, of subsequent

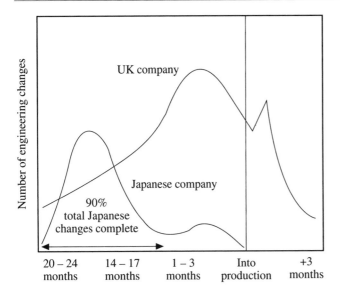

Fig. 1.3 Engineering change—volume and timing comparison

changes further down the line. This concept is shown briefly in Fig. 1.3 and discussed in more detail in Section 1.4.

1.3 The pace and influence of technology developments

We are all aware that today's manufacturing and industrial enterprises, in line with all other business environments, are having to be aware of, and implement where applicable, an increasing range of advanced technologies and philosophies in order to remain competitive (Fig. 1.4).

These individual areas such as CAD, CAE, FMS, CAM, MRP, etc., are generally well documented and understood but epitomize a trend which has affected us all over the past number of years—the acceptance of the TLA.

A few years ago I attended a technical seminar and the after-dinner speaker began talking about TLAs. The effect was interesting—almost everyone in the audience (including me) didn't have a clue what a TLA was yet no-one was prepared to take the initiative and ask. After a while the speaker had obviously achieved his aim and enlightened his audience: a TLA is a Three Letter Acronym such as those mentioned above! His point was to highlight the fact that we all now accept these acronyms as part of our everyday speech and rarely question the justification and meaning of them or their use. They centre everyone's attention on the technicalities of each individual one and in general we are afraid to question them for fear of appearing ignorant. Indeed, the use of TLAs has become so common that certain groups of people, especially computer specialists, use them almost as an everyday language.

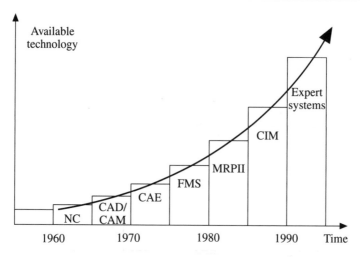

Fig. 1.4 The exponential trend of technology development

However, their use is not all negative. It is true to say that without them we would require to repeat long verbose descriptions of philosophies or techniques time and time again to communicate with each other. What we require to do is to 'de-mystify' TLAs and remember that they are only descriptions of techniques or tools which we can use to enable business benefits to be achieved.

This concentration on technology for its own sake can also be seen by looking back over the past few years. The collective growth of individual technologies has increased in an exponential manner and, although a number of individual transition 'discontinuities' have occurred, the overall rate of increase shows no signs of slowing over the coming years as we have already indicated in Fig. 1.4. In many cases the adoption of these tools was a direct result of the driving strategy of many companies in the 1970s and 1980s to become the 'lowest cost producer', where the emphasis was directed to improving individual or departmental productivity. Keeping abreast of all these techniques is no easy job and most engineering/manufacturing companies now employ at least one computer applications 'guru' to attempt to maintain a watchful eye over current and emerging technologies. Consultants are also often used periodically to review a company's operations and suggest areas of improvement in this respect.

Many will remember the 1980s in particular as the decade of information technology (IT) where the emphasis was placed on the computer and its associated techniques and software. This was exemplified in the UK by the Government creating a 'Minister for Information Technology' post to help companies (via awareness training, grants, consultancy assistance, etc.)

understand and introduce various forms of computer-based assistance. It is always easy with hindsight but although this was a welcome initiative, it tended to concentrate on technology for technology's sake without consideration of the business-related benefits to be gained. If the techniques cannot be justified then one should not be drawn into using them unnecessarily.

EDM and certain other similar concepts within a computer-integrated manufacturing (CIM) framework operate over multiple departments and divisions of a company. They should be considered as 'unifying' technologies rather than 'point' type solutions such as CAD or CFD. These individual solutions can prove to be very effective in their own field of operation but very often only succeed in moving the resource bottleneck to another area. For example, a company which finds it is not able to respond to its chosen marketplace quickly enough may have identified the drawing office as being the resource bottleneck in generating drawings for production use. So it invests in a CAD system and, sure enough, the output of the drawing office improves— but it still cannot meet market demand. Further studies show that the production of technical documentation is now the cause of the problem and it invests in a desktop publishing (DTP) system. Now the bottleneck could move onto the design or manufacturing areas, and so on. Each company's IT requirements differ and each one must choose its own 'set' of tools which address not only bottlenecks but also the overall management and control aspects.

Although EDM as a concept has been around for some time now, recent advances in related areas of technology have moved it along considerably. This is often the case with advances in technology (both via research and later practical application)—a synergy of related technologies occurs where the whole is greater than the sum of the individual elements and a 'coming together' or convergence in them occurs. For example, current EDM systems could not function as they do without the related advances made in database technology, or networking and communications, or indeed user interface technology.

There are two main sources which drive the pace of technology:

- The users of existing computer software tools who initiate technical developments by various means. One example could be the user of a product who requests a particular improvement to the existing software. After considerable discussion and lobbying this may be included in the next release by the software supplier. After a time this functionality is seen as the 'norm' and other requests are made, and so on.
- The research establishments and solution suppliers who either develop a new tool or methodology and set out to create a market for it or who may need to extend an existing product for various marketing reasons.

Both of these sources fuel a 'spiral of expectancy' (shown diagrammatically in Fig. 1.5) relating to computer software products which not only increases the

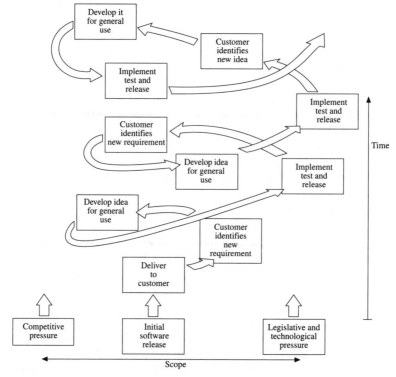

Fig. 1.5 The computer software related 'spiral of expectancy'

pace of development but, if not properly controlled, can also quickly develop into an uncoordinated and unstable situation.

So, we require to learn from our past experiences relating to the rise in technological developments and harness this for future actions. We need to remember that 'information', which a computer is good at storing, is not 'knowledge' and use computer system output wisely and for practical use. We need to remember that there can be a lot of hype in relation to new computer and other technologies and this has to be cut through to get to the basic facts (remember the hype surrounding MAP and TOP protocols?). We need to remember to apply computer tools in a collective sense—to take a wider view, not a short-term blinkered approach. We need to remember and enjoy the excitement of applying new technologies to solve our business problems and make them work.

By utilizing and learning from our experiences we should be better able to respond to the challenges of the 1990s. Current thinking suggests that these challenges will closely follow the principles involved in 'concurrent' or ' simultaneous' engineering, i.e. to improve and streamline the process:

● Do it right first time.

- Make increasing use of teamwork and cooperative thinking.
- Involve the use of product teams.

These are concepts which we will consider in more detail later.

1.4 Manufacturing excellence—the need to embrace new technologies

The challenge facing industry in the 1990s and beyond is to connect people, not just technologies, together. We require to get back to basics—to understand the real business issues which can help us grow—to understand the *issues* not the *three-letter acronyms*. In many cases this requires us to take our blinkers off and look around—to take a more strategic view of our problems and to strive for solutions to them in a global sense. EDM for example is sometimes known as a 'distressed purchase' because the company concerned has perhaps begun to have major documentation problems, which are probably beginning to affect its customers, and it is forced into taking action in a reactive, rather than proactive, manner. EDM, because of its potential long-term effects, requires to be considered as part of a strategic plan. Within these overall views we still need to incorporate those parts of technology which best fit our individual business needs. This process is called integration. The result is business excellence. Business excellence requires all areas of a business concern—design, manufacturing, sales and service, etc.—to excel both individually and collectively to achieve 'world-class' status.

EDM and other related technologies go together to form component parts of a larger picture but very often we require to *think higher and wider* in order to visualize this picture before we make a start on one small element. Other related philosophies and technologies also form fundamental parts of the route towards business excellence or world-class manufacture, for example TQM, JIT, CE. These will be discussed later in this chapter.

World-class manufacture and business excellence are phrases which roll off the tongue very easily and can readily be identified with. But what do they mean in practical terms? For example, with regard to EDM, the point to remember is that *EDM is a solution, NOT a product*, and that introducing it in isolation will not achieve the full benefits which are obtainable if considered as part of a strategic plan. As Andy Coldrick, MD of Oliver Wight UK Ltd, quotes, 'Many senior managers do not feel that it is their role to ensure that initiatives to improve the business are implemented. When they accept the connection between strategic planning, tactical planning, and execution, and the use of quality functional deployment to enable them to go through a strategic planning process that is more customer focused, they will become the leading companies.'

So how can business or manufacturing excellence be defined? In simple terms it means having the right manufacturing capability to profit from totally satisfying the customer, with high quality service and products at the right price, delivered at the right time. It is a fundamental cornerstone on which to succeed in your chosen marketplace and depends heavily on other appropriate techniques and philosophies, including JIT, MRPII, TQM, FMS, and many more.

A recent study by Cardiff Business School, Cambridge University and Andersen Consulting confirms the need for world-class thinking in order to improve our global standing. The study concludes that the best world-class Japanese manufacturing plants outperform their UK equivalents by at least 2:1 in productivity, throughput time and the use of space, and by 100:1 in terms of quality. It also shows that not all Japanese companies are world class—many still have a long way to go to achieve this status. However, the gap between world-class and non-world-class performers is large, with half of the UK companies falling in the lowest performance quartile (Fig. 1.6).

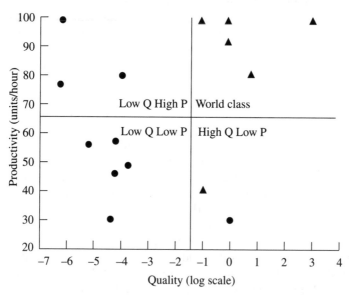

Fig. 1.6   World-class company matrix (*courtesy of Andersen Consulting*)
▲ Japan, ● UK

This would seem to suggest that there is considerable scope for improvements in UK manufacturing, especially since industry volumes are expected to increase dramatically in the next five years or so. Small incremental changes will not be enough to succeed: quantum leaps are required.

So are you (or are you thinking) 'world class'? The DTI's Enterprise Initiative *Managing into the 90s* quotes some of the challenges that potential world class manufacturers should address (as defined by Professor Colin New of Cranfield School of Management). They should look to:

- Reduce inventory investment by 50 per cent or more.
- Reduce manufacturing lead times by 50 per cent or more.
- Introduce new products at 2 or 3 times the present rate and at 50 per cent of the present new product lead times.
- Reduce manufacturing costs by 30 per cent or more.
- Reduce overhead/support labour by 50 per cent or more.
- Improve quality to a 'parts per million' defect level.

Achieving world-class status involves a number of steps, a 'methodology' which requires to be applied to define and achieve objectives. These are:

1 *Get back to business basics*   By deciding what it is that you wish to excel at.
2 *Identify the key elements*   To achieve your aim defined above.
3 *Understand your competition*   Using tools such as the competitive positioning analysis chart shown in Fig. 1.7 to profile your performance.
4 *Understand your customers' needs.*

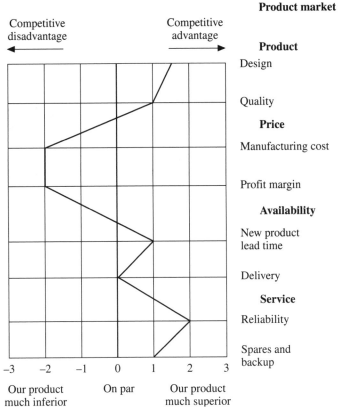

Fig. 1.7 Competitive positioning analysis (*courtesy of Cranfield School of Management*)

5 *Select your tools for success* Including products, services, equipment, people, information and systems.

Forward planning is the key to world-class manufacturing and business excellence—without it you will not succeed.

There are many other pressures exerted on a manufacturing business, both of an external and internal nature. Responding to these in a timely manner is a further important area. Let us summarize a number of the external driving forces, some of which have already been considered:

- Increased competition
- Increasing use of standards—from many sources. e.g. CALS, ISO 9000
- Expanding multicompany organizational structures—perhaps as a result of take-overs or mergers
- Shrinking internal budgets—for both DP/IT and general capital goods
- Increased automation capabilities
- The need to maximize (leverage) existing computer investments
- Economic recession

Many companies respond to these types of pressures by forming internal objectives such as those typified in Fig. 1.8. To do this requires that philosophies, not just technologies, are put in place. It requires a recognition that information technology is only a tool to perform a job and that often it is change—be it at a strategic, cultural or procedural level, tied to appropriate tactical planning and control—that is required to effect the improvements on the scale required.

The key processes involved under the banner of business excellence or world-class manufacture are as follows. It is important to remember, however,

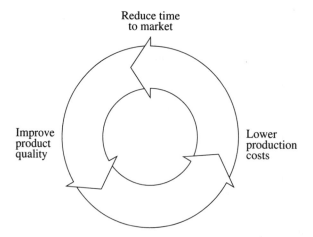

Fig. 1.8 Internal objectives to respond to market pressures

that it is not the adoption of such processes in isolation which forms the key to success. It is quality, reliability, delivery, flexibility, cost and, most importantly, providing your customers with better value than your competitors. It is also often wise to introduce such processes in a sequential manner—running a number of large projects simultaneously can often lead to the 'change machine' getting out of control and not delivering the expected returns.

MRPII (Manufacturing Resource Planning)

This is a planning and control system which has grown out of Material Requirements Planning (MRP). MRP's central tenet is to provide a planning tool to assist in the calculation of the correct amount of material and resource needed to produce the required output. MRPII can be defined as a priority planning system. It has developed from the practical use of available tools and extends the concept of closed-loop MRP into a planning tool for all the resources of a manufacturing company—Manufacturing, Sales and Marketing, Finance and Engineering. MRPII is literally a simulation of a manufacturing business. It can be used to schedule the factory, schedule suppliers (vendors), plan manpower requirements more efficiently, plan capacity requirements more accurately, can be tied into the business plan and used to simulate 'what-if' situations. MRPII is being taken into the central concept of other acronyms such as Enterprise Resource Planning (ERP) and Customer Oriented Manufacturing Management Systems (COMMS). These are defined as sets of technical requirements for the 'next-generation' manufacturing systems and differ from MRPII because they look beyond manufacturing to other functions in an organization. The essence of ERP for example is looking forward to the integration of the entire organization. This would appear to be the same goal as CIM but without explicit mention of that dreaded word 'computer'. ERP would, of course, be very difficult without one, in any sizeable business enterprise.

TQM (Total Quality Management)

This is an approach to Quality Management and Control which blossomed in Japan, but was the result of a series of lectures in Japan in the 1950s by two American consultants (W. Edwards Deming and J. M. Juran). It implies a new style of thinking in most companies, i.e. where everyone plays a part in achieving the 'perfect' product. It is an essential element of CIM and requires the use of a new level of technology in measurement, test, inspection and gauging as well as the associated cultural change. TQM itself utilizes Total Quality Control (TQC), the underlying philosophy here being that any physical defects are considered unacceptable and that effort must be directed at continuous improvements to achieve this 'zero defect' state.

This contrasts with the approach in much of Western industry which says that there is an acceptable quality level (AQL)—a certain percentage of defectives which it is considered uneconomical to reduce further. TQM embraces many areas of a company's operations such as attitudes, problem-solving tools, teambuilding, quality of information, quality of management and the need for continuous improvement—the belief that we can never be so good that we cannot get better.

JIT (Just In Time)

This is also known as zero inventories (ZI) and 'stockless production' and is more of a philosophy than an 'off-the-shelf' tool. Its goal is to eliminate all waste in the entire production cycle—minimizing material, manpower, machines or tools necessary for production. In particular JIT aims to produce in minimum lead time and provide just the correct amount of material (at the specified quality) just when it is needed at each stage in the entire production process. JIT was developed over a 15 year period by Toyota Motors in Japan. The ideal JIT system has zero defects, zero inventory, zero lead time and a batch size of one—although unobtainable the aim is to drive towards these goals.

TQC is generally regarded as a prerequisite to implementing JIT. Since inventories are reduced to a minimum so that quality problems are not buffered by inventories, it follows that any quality problems which are not virtually eliminated could stop production.

Change management

This is sometimes considered as part of TQM but can be broken out as the acceptance of new techniques and methods of working. The context in which 'change management' is used here should not be confused with the engineering change management/change control process which is a procedure used to control a product's configuration/structure/history and is, as we have already indicated, a fundamental part of EDM. The use of the term change management, or business process re-engineering (BPR) as it is more commonly known today, refers to changing the cultural aspects of perhaps long-established practices and working procedures. Motives for complex change come from all sorts of business forces such as mergers, expansion, contraction and so on. But it always involves a restructuring programme across one or several departments and usually has a big impact on the workforce. It is easy to underestimate how painful change will be for a company and also how difficult it is to stay on track and achieve your original goals. The aspects of BPR and the change management ethos were recently summarized by one of my colleagues. He stated, 'Change is an opportunity, not a problem—we must strive to think outside the box rather than stay in it.'

Concurrent/simultaneous engineering

Cutting the time it takes to develop products has an undeniable business logic. The question we have to answer is: how? Organizational and entrenched 'mind-set' change is important since CE involves teamworking concepts (to the highest level) which can initially be difficult for a formally-structured company or a serially-minded designer to come to terms with (as line management functions become blurred)—but the message and the technique are getting through. Various recent studies show that up to around 80 per cent of manufacturing companies are now using multidisciplinary teams to develop new products. But, traditionally, cutting time to market means trying to solve as many problems as possible early in the design cycle—a Design for Manufacture and Assembly (DFMA) approach as shown in Fig. 1.9.

This process may or may not involve computer applications: the savings do not necessarily come from the individual applications or techniques themselves but more as a result of how they are implemented. The aim of the CE process (which incorporates DFMA) is, of course, to remove the concept of 'over-the-wall' engineering principles where industrial engineering/manufacturing requirements are only considered after the design and detailing phases are complete. An overlap must be generated here and the larger the overlap the more teamworking aspects come into play. This is well represented by Barry Brooks of PA Consulting Group who refers to the 'Brooks wedge' to clarify this concept (Fig. 1.10).

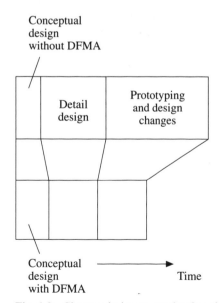

Fig. 1.9 Shorter design to production times through the use of DFMA early in the design process (*reproduced courtesy of Boothroyd Dewhurst Inc.*)

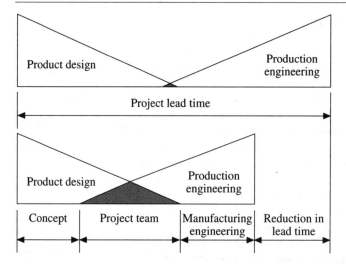

Fig. 1.10   The Brooks wedge (*courtesy of PA Consulting Group*)

The other main benefit in adopting CE principles apart from reduced time to market is of course the control of product costs. This was mentioned briefly in Section 1.2 and is further explained with reference to Fig. 1.11. The main reason that costs are affected in this way is that engineering changes to the product caused by incompatibility between the design and the manufacturing process inevitably result in problems delivering the products to schedule. The key to concurrent engineering is early validation of product requirements by all 'consumers' of the design—the adoption of a 'right first time' approach. This early participation by all parties in the supply chain, which extends from

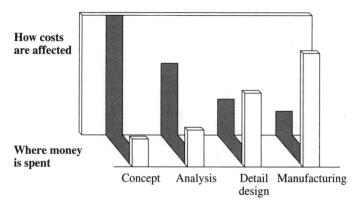

Fig. 1.11   Design process cost commitment

the supplier to the customer, helps to identify design flaws that may prevent the successful manufacture, test, marketing, or support of the product. Most companies actually spend the majority of their product development moneys during the production planning and actual production phases of the product's life-cycle. But since the product design dictates manufacturing methods and techniques, most of the overall cost of manufacturing a product—studies indicate this to be around 80 per cent—is actually committed during the early design stages, for example production processes, machine requirements or perhaps complex casting or moulding requirements. What is required is to consider the implications of downstream activities much earlier in a product's development—'front loading' the whole process with more of a 'Design for Manufacture' approach. Although the cost benefits are not immediately apparent, increased revenues and an earlier break-even are definitely there for the taking and present the best overall approach to modern product development methodologies. As can be seen, the concepts in concurrent engineering and EDM are very closely aligned—tackling one generally means addressing the other.

2

EDM—its place within an integrated environment

2.1 CIM—its definition and constituent elements

You may not believe that there is much of a connection between the computer applications industry and the fashion industry, but they are very alike in many respects. Each one of them has a very cyclic nature with particular trends being popular or unpopular at certain times and long cycle times between the highs of popularity and the lows of depression.

The clothing or fashion industry of course will have seasonal cycles super-imposed on this overall trend which the computer solutions industry should not! Each one has the forward looking younger element who are keen to adopt new ideas and want to lead the pack. Each one has those who may take a more seasoned view and conform to the generally accepted view of things—the grey suit owner as opposed to the bright purple suit owner. And each one has the 'steeped in tradition' brigade who find it hard or don't want to move along with the times—for example the person wearing the pin-striped suit.

In the computer industry there are new solutions and computer aids appearing on the market every day. Some are adopted quickly and/or prove to be fundamental building blocks for others to follow (for example MRP). Others are fashionable but are quick to die, prove unsuitable or are superseded by another technology (who backed the Betamax VCR tape format?). Trends and fashion play a part in both areas with philosophies/technologies/tools and methodologies constantly coming and going in the computer/systems environment in a similar manner to the jewellery/cloths/cuts and colours which are constantly changing in the fashion environment.

One of these in the systems area is Computer-Integrated Manufacturing (CIM) which many people now feel is dead or 'old hat', but which actually still underlies many of our current endeavours in refining and integrating our individual disparate computer systems into a cohesive and structured unit. So the TLA is unfashionable but the requirement is not. Who do you know in today's business environment who purchases an expensive computer system without considering whether it is compatible or integrates with what already exists? EDM and CIM are closely aligned—each one is concerned with the control and management aspects relating to systems and data so we shall spend some time considering CIM here to put EDM in perspective.

Because it is so diverse, CIM has many definitions. However, the central concept is that a successful implementation will result in a manufacturing business where all processes are integrated and balanced—from design of the product to its delivery to the customer and beyond—i.e. where production planners, supervisors and accountants share the same data as designers and engineers in order to plan, execute and control all activities within a manufacturing enterprise. But it should always be remembered that the focus must be on business integration, not just computer integration for its own sake. Perhaps a new acronym should be established to convey its proper meaning— CIB: Computer Integrated Business! Indeed the Gartner Group, a leading market research company, has developed the concept of CAPE (concurrent art to product environment) as a fundamental aspect of an integrated development environment. CAPE highlights the need for a new set of integrated, enterprise-wide technologies and processes to support a design and product environment. It shares the CIM view that corporate strategies, processes and technologies require to be balanced, and recognizes the need to embrace concepts such as CE and technologies such as EDM. The concept of CIM is shown diagrammatically in Fig. 2.1.

Ideally CIM develops coordination among business functions and helps to refine, modify or reorganize some functions to meet customer needs and competitive pressures. Like most topics of a strategic nature, CIM is not a product

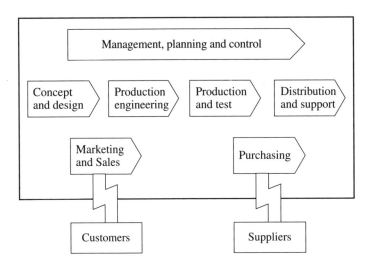

Fig. 2.1   The CIM concept (courtesy of PA Consulting Group)

you can rush out and buy for immediate installation. Rather it is a strategic philosophical approach that takes several years of development. It begins with a top-management level understanding of future requirements and target markets and employs the most up-to-date technologies (which may be changed or added-to constantly—hence our interest in EDM as a part of a CIM environment). CIM can therefore be thought of as a 'toolkit' of techniques and technologies applied within a clear strategic framework. Some of these tools are available off-the-shelf (e.g. MRP, CAD/CAM) and others are essentially philosophies themselves (e.g. TQM, JIT).

The application of such techniques on their own can create 'islands of automation': CIM's role is to integrate the whole range of manufacturing and related activities to avoid this situation, hence the term C*Integrated*M. In the past when CIM was a growing concept there was considerable discussion as to where the emphasis should be placed. For example, the DP/IT department may take the *C*im view where the *computer* is the driving element, or the production/manufacturing viewpoint of ci*M*. Time has proved the real value to an organization of CIM being c*I*m where the emphasis is firmly placed on *integrating* these 'islands of automation'. A range of CIM techniques and their relative operational areas is shown in Fig. 2.2.

Like most 'sizeable' projects and solutions, there are disaster stories of failed CIM implementations, in a similar vein to MRPII and CAE implementations. The most frequent reasons for these seem to be either poor timing,

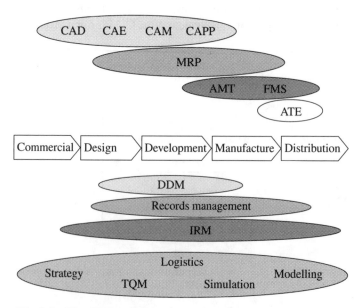

Fig. 2.2 The CIM spectrum (*courtesy of PA Consulting Group*)

poor selection of technologies to be used or a lack of commitment. Poor timing can take two main forms. The first involves investing in unproven technologies. The second arises from inertia—moving too late when perhaps the competition is already well down the road. The first is less of a problem in today's environments where there is sufficient technology available to make implementing CIM worthwhile for most manufacturing sectors, but it is still worth taking care here—do not go too much out on a limb. The second point can be much more dangerous. Given that it can take years to implement an integrated computing environment, companies that only respond when the concept is proven are likely to find themselves in an untenable position.

The benefits in adopting an integrated computing strategy will be briefly outlined in the next section of this chapter and, when implemented correctly, can be dramatic in nature. For example:

- Increased machine utilization
- Reduced labour costs
- Reduced lead times
- Reduced WIP inventory
- Virtual elimination of set-up time
- Flexibility in changing schedules and product mix

However, success in computer-integrated manufacturing requires that a company has a long-term strategic business plan and implements and integrates each chosen element in a clear logical sequence. Now that the concept of CIM is well and truly established, any future advances in business and manufacturing technology coupled with those in computer hardware and software can all be channelled into the concept as and when required.

2.2 The strategic nature of CIM

A number of years ago, when I was employed by a UK supplier of CAE systems, the concept of CIM first appeared and everybody started to jump on this particular bandwagon as it went by—advertisements appeared for CIM 'systems' and we would receive phone calls from prospective clients asking for details: 'Can I buy one of your CIM systems please?'

Well, the answer is of course 'No' because CIM is a long-term business strategy, not a product. It is an ever-expanding and changing concept, so it is unlikely that a 'finished' or 'complete' implementation will be found. Truly integrated manufacturing must be achieved rather than bought and this concept is just as applicable today as it was when the philosophy was introduced.

Like most items of a strategic nature, success depends very much on leadership and commitment. Commitment from the very top of an organization is vital because one of the first steps in detailing an integrated strategy is to take stock of the current situation and define a vision of the future—a strategic three to five year business plan. Part of this plan will involve the decision on what automation and computer aids will be required. Defining a strategic business plan without consideration of upcoming technology or vice versa will prove fatal to the success of both.

What is required is a detailed implementation plan to achieve this overall aim: a CIM strategy which is in line with the company's long-term aims and assists with the desired performance objectives. This is shown diagrammatically in Fig. 2.3.

CIM is not something which can be implemented with a 'big bang' approach. It will take at least 12 to 18 months to achieve initial results from the first stage and probably 5 or 10 years before it can be said to be reasonably established. So it should be planned as a series of individual projects, each one of which brings its own benefits, while all contributing to a coherent strategy.

Looking at each area in isolation without this overall view is a 'non-optimal' route which has already been experienced by a large number of companies in the technology-focused and 'cost-centred' approach of the 1970s and 1980s. A short-term view of computerization results in bottlenecks sliding up and down a company's chain of operations, as we have already discussed in Chapter 1. A longer term strategic or enterprise view aligns each part of the jigsaw (such as CAD, DTP, EDM or JIT) into a larger cohesive picture. In essence it is a computerized 'shadowing' of the business excellence philosophy we have previously outlined.

So what are the benefits of a CIM strategy? Why if it is such an important aspect of a company's operations is it not a TLA in constant use, read about in trade magazines and discussed at exhibition seminars? Well it is, but usually

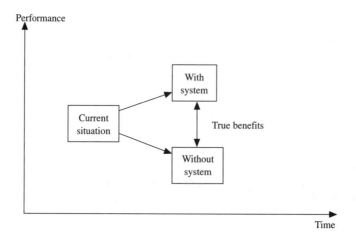

Fig. 2.3   The benefit potential of an integrated computing environment

under a different name: 'Integrated computing strategies for the 1990s' and 'IT for the manufacturing sector' are two examples which spring to mind. The benefits can be immense but are very difficult to measure—and I make no apologies for the fact that the following benefits are very similar to those discussed for EDM systems. As we have seen previously, there is an increasing trend for many of these tools to come together. They are 'unifying' technologies which are now driving towards satisfying an increasingly standard set of objectives which manufacturing and business enterprises see as being fundamental if our traditional manufacturing industries are to be preserved.

EDM is a relative newcomer to the CIM arena but already the trend is becoming clear: computer-integrated manufacturing is difficult, if not impossible, without EDM. The different demands of technology and markets among manufacturing industries generally mean that each organization identifies different primary benefits from investing in CIM. However, there are a number of common benefits which can be identified. These are, briefly:

- *Flexibility*   The need to adapt to changes in the marketplace, technology advances and perhaps product mix variables (if one production facility is being used to produce a number of standard products) is increasingly required. Gone are the days where the automatic choice was to install a large dedicated production machine. Although these are able to produce at lower unit cost than smaller, more flexible machines, their efficiency stems from the fact that there will be minimal change in the product and a relatively high stable demand—factors which nowadays are considered a luxury rather than the norm! In addition, the cost of changing the production line can also prove prohibitive—a factor which Toyota Motors of Japan took into account when establishing their SMED (single minute exchange of dies) philosophy.

- *Reduced costs*   Reducing unnecessary costs is a common goal among all enterprises in these difficult times and the application of an integrated computing strategy can also help here in many ways, for example in controlling production costs by better application of advanced manufacturing technology (AMT) such as CNC machine tools or flexible manufacturing systems (FMS), and by better production scheduling by the correct adoption of line balancing techniques or MRP/MRPII. Major cost savings are possible through the correct application of inventory control and material management techniques such as those provided by MRPII and JIT which have cut unnecessary inventory holding by as much as 50 per cent.

- *Increased management information*   One of the most valuable benefits that CIM brings with it is information, but the point to remember here is that information is just that—information. It has to be relevant and applied correctly if it is to provide the main benefit usually considered when data is mentioned: that of increased *control of the business*.

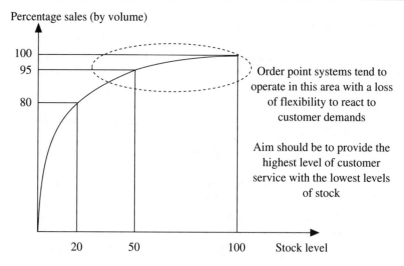

Percentage sales (by volume)

Order point systems tend to operate in this area with a loss of flexibility to react to customer demands

Aim should be to provide the highest level of customer service with the lowest levels of stock

Fig. 2.4   Customer service versus stock

- *Reduced order lead time*   Today, the manufacturer who can provide a faster and more reliable product delivery has a major competitive advantage. The traditional method of doing this was of course to hold the maximum amount of finished stock (Fig. 2.4). The manufacturing industry can no longer afford this luxury and is now required to find alternative methods, such as improved production scheduling tools as found when moving from the traditional order-point based systems to MRPII, or perhaps via EDI.

  A customer-driven manufacturer will prove to be a winner in times of strong competition and a comprehensive integrated IT implementation is required to support this philosophy.

- *Increased customer satisfaction*   How can increased customer service be defined and measured? Well, it relies on a quality service from start to finish—ranging from a speedier supply of the required product at a competitive price, through a complete after-sales service (including warranty and maintenance contracts), and on to the supply of updated or new products.

- *Reducing 'time to market'*   The days of the traditional product life-cycle approach where a new product is introduced as an older one is in decline are no longer typical of today's manufacturing environments, especially in faster moving markets such as computer hardware or consumer electronics. Companies in these sectors have to have the 'next generation' product in production planning and maybe even others up and coming to back them up. The majority of products are getting more complex and take longer to get to market, but at the same time their life-span is reducing. There is a real danger that businesses cannot respond to this challenge. However the realities of this situation now appear to be understood and accepted. For example, the 1993 Computervision Manufacturing Attitudes survey showed that 72 per cent of respondents agreed that 'design to manufacture' was a key area to address, and action is being taken to address these issues with EDM forming a core element of many CE environments.

- *Increased product quality* The concept of TQM has already been discussed in the previous chapter and it requires the backup of an integrated computer system to make it work. CIM's role in a TQM environment is potentially very wide and can include areas such as:
  - enabling engineers to perform the job they were employed to do rather than chase paperwork and find information
  - avoiding the pitfalls associated with data transcription errors which can prove a serious source of error with un-integrated solutions
  - providing better control and planning systems such as those enabled by MRPII, statistical process control (SPC), etc.
- *Better use of people* This aspect is again considered as part of the TQM process where respect for working colleagues and employees is considered a necessary part of the overall environment and manufacturing process. The correct application of CIM can enhance the creative skills of the persons involved, providing increased job satisfaction and act as a key means of retaining good staff.
- *Meeting the increasing number of external requirements* Manufacturing industry today is labouring under a steadily increasing number of safety, environmental and legislative requirements. In order to incorporate these into everyday use some form of computer assistance is usually required and again an integrated approach can assist in providing a suitable framework to accommodate these requirements.

2.3 The driving forces behind the emergence of EDM

The convergence of technologies and computer-assisted tools to enable us to perform our daily tasks is ideally illustrated by EDM. As we have discussed, these tools and computer application systems are all now driving towards similar goals and business benefits. This is not to say that the individual 'point' solutions such as CAD, DTP, MRP, etc. do not have a place—they are important elements of a much larger picture and in general are now reasonably well known and used. But the idea of CIM, the concept of integrating the computerized elements of a company's operations, cannot be fully realized without EDM. It is *data* that forms the central concept of the entire operation: its processing, integration, sharing, extending, enquiring and reporting. Indeed, it is a well-known fact that the value of data to a company after a few months of operation far outweighs the capital cost of the equipment on which it runs— yet many people still concentrate on cost savings and management of the hardware and pay scant regard to adequate control and management of the data itself. Hence the usual situation of considering disaster recovery only after a serious fire where perhaps it is too late to recover. Useful data is usually referred to as information, and today's more enlightened IT customers are pressing their suppliers into providing more and more powerful tools to translate data into this most useful of commodities. It is important, however, that information is presented at the correct level. Although we need information for correct decision making and action, the presentation of too many irrelevant facts and documents has the same effect as too little. The process of generating and disseminating information is generally referred to as information management.

The control of this information is of paramount importance and since the vast majority of downstream data generators are actually located in the product design and development areas it follows that the control should start as early as possible in these areas and continue to be used for as long as individual requirements dictate. If the management of the associated data is poor, the end result is likely to be poor product quality, high costs, errors that are discovered too late, delays, etc.—all leading to possible lost future orders. It therefore follows that the forces which drive the acceptance of CIM also drive EDM, albeit at a higher level. These topics, such as the current recessionary nature of the world's economy and the shift in the balance of the world's manufacturing centres, have already been mentioned. But there are others which, although they may be at a lower level, are just as important. These include:

- *Increased competition*   Unless you are in the fortunate position of having a unique market sector all to yourself you cannot have escaped noticing the cut-throat nature of competition in the majority of today's market sectors. Anything goes in selling more than your competition nowadays, not just a better or more advanced product, but direct marketing initiatives, distribution and joint marketing agreements, price, special incentive offers (who could forget the 'free flights' offer by Hoover) and so on.
- *Increasing product complexity*   Technical advances in product design are subject to similar 'expectancy spirals' as computer software solutions, where the marketplace expectancy is constantly rising and the first supplier who meets these expectations can obtain a significant market share over their rivals. The rapidly expanding market for hand-held camcorders is a prime example here. The move from mechanical to electronic and computer-based products also plays a significant role—for example, railway signalling and telecommunications. Software revisions etc. require to be managed correctly, as do the increasing number of components found in modern products compared with their older counterparts.
- *Time to market*   When the above two factors combine, a dangerous situation can occur. This is where a company cannot develop products at a fast enough rate to keep up with market demand (Fig. 2.5). Where are your

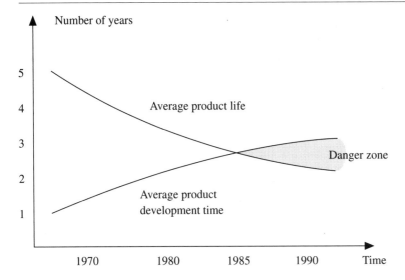

Fig. 2.5   Product development timescales

company's products positioned on this diagram? If they are nearing the 'danger zone' what plans do you have in place to address the problem?

- *Standards*   The increasing number of standards both of an enforced *de jure* and a *de facto* nature are playing an important role in ensuring a successful product introduction. The effect of these standards (such as ISO 9000) and the need to comply with them cannot now be ignored. Defence-related initiatives have also increased the level of awareness of EDM within all manufacturing environments. Standards have been developed such as CALS (Continuous Acquisition and Life-cycle Support), which aims to reduce the level of paper-based data transfers in the areas of product data, support data and technical manuals. CALS was originated by the US DoD. It impacts heavily on DoD suppliers initially but, like most US standards, eventually affects the UK MoD and subsequently manufacturing industry in general. This situation is already apparent with rising interest levels in EDI and CALS type technologies/standards.
- *Increased legislation*   The increase in government controls and liability laws associated with safety and environmental issues, resulting from poor quality product design, development and subsequent use, all impact on current manufacturing businesses. Although traditionally centred on the aerospace industry, these aspects are becoming increasingly expected of automotive and all other product manufacturing environments.
- *Teamworking*   In order to respond to market pressures and continue to ensure market share, many companies are tackling areas other than just technology to provide them with an edge over their competition. Many

have changed their entire management structures to enable faster reaction to relevant demands, empowered junior management to take action when required and adopted a teamwork or 'workgroup' approach to product design and development. Team/group working requires that the relevant tools and resources are at hand to enable the benefits of a 'fast track' approach to be achieved, and this includes the access to relevant data, normally via suitable groupware applications (see below).

- *Increased computer automation* Computer technology provides the enabling tools for many of the generally accepted techniques we currently employ. This sounds obvious but the effects of hardware, processing, network and database advances (which are all areas of rapid development) are very often overlooked when considering the associated application software. These technologies, especially networking, distributed processing and groupware applications such as electronic mail, diary systems and information sharing and scheduling tools, have (and will continue to have) a significant effect on the ease with which EDM systems (which may have to interface with many files on a range of hardware) are developed, installed and maintained.

- *Shrinking budgets*   It is highly unlikely that your company's DP manager is going to get the £250 000 budget for that new computer system that would probably have been more forthcoming a number of years ago. Money, as we all know, is very tight—people require to make maximum use of what they have already got. EDM can provide a lower cost investment than some larger, more prominent, projects and return significant benefits as we shall see.

- *Maximization of existing computer investments*   The localized or point solutions referred to earlier can sometimes make the situation even worse:
    - The use of digital data has exploded in recent years with the majority of companies having large pools of digitally held data. These are usually files from diverse systems and cannot be readily re-used.
    - Thus 'information islands' have been created where inadequate control over the data results. Rather than being generated and left on a backup tape because it cannot be found easily for re-use, many companies now want to actively use this information for update purposes: it is more cost effective to re-use data rather than regenerate it. This also applies to the data belonging to any older technology solutions—often termed 'legacy' or 'heritage' systems.
    - In many cases some sophisticated tools do exist to help in specific automated procedures, for example a CAD system which can assist in the automation of drawing changes. Such tools facilitate changes in perhaps an uncoordinated and localized manner. Excessive design changes are generated and can cause bottlenecks further downstream, i.e. the entire process goes out of control.

These 'localized automation problems' were summarized well by Ed Miller, President of CIMdata, a consultancy company specializing in EDM, during a recent UK presentation. He stated, 'The speed and capacity of certain computer tools exceed the management controls that can be exercised over them.'

We require to regain control over these tools by improving our methods of working, by utilizing an approach to product development that emphasizes up-front involvement from all affected members of the process. This process is *concurrent engineering* (CE) and was introduced in Chapter 1. Concurrent engineering is the interactive collaboration of all specialists early in the overall process, thereby evaluating all product design parameters initially, and avoiding the iterative cycles characterized by the more traditional sequential product development approach. Concurrent engineering is also known by many other names such as 'design for production' and 'right first time'. It is, however, important to recognize that CE is more of a management philosophy which entails a fundamental re-appraisal of all manufacturing operations rather than simply making computers communicate better.

CE is not an easy concept to introduce. This is confirmed by the results of a recent study which showed that less than half of UK manufacturing companies actually practised concurrent engineering principles. Further, its use is low in the small-to-medium enterprise (SME) sector, believed to be vital to the regeneration of the manufacturing base in the UK. However, it is being increasingly recognized as one of the major differentiators a company can have in the 1990s to maintain its competitive positioning in its chosen marketplace and many engineering and manufacturing companies are now showing remarkably high levels of awareness and knowledge of the need for CE. It addresses the vital concept of reducing a product's 'time to market' and relies on EDM to enable the data 'concurrency' to exist in conjunction with the physical, cultural and communication infrastructures required to enable the successful adoption of CE principles. It is widely accepted that a product late to market can lead to a 30 per cent loss of profit potential and that product costs 10 per cent over target can lead to 20 per cent less profit potential, both common situations. Getting it right first time can reduce, or perhaps even avoid, such problems.

Apart from shorter development timescales, other benefits which are already being achieved from the successful use of CE include reduced product costs, shorter manufacturing times, improved product quality, reduced operational costs and increased manufacturing flexibility and market share. Many companies which cannot currently effectively share data across the engineering group may tremble at the thought of sharing ideas and information in an integrated process that involves design, purchasing, manufacturing, marketing and other organizations, but through a combination of changes in company culture and tools, the roadblocks that stand in the way of a successful CE implementation can be removed. Figure 2.6 shows the typical obstacles that require to be overcome to successfully benefit from concurrent engineering.

Percentage of sites

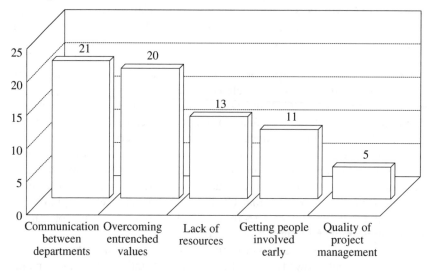

Base: All sites aware of concurrent engineering (119) / Spontaneous

Fig. 2.6 Issues faced when implementing concurrent engineering (*source: Computervision*)

2.4 The role of EDM in an engineering enterprise

It has been stated many times that engineers do not make products, they produce data. This of course is not their intent, but it does arise in a number of ways. For example, you may remember a situation where you have needed to develop a particular idea into something which is going to cost a fair amount. It may be the generation of a new jig or fixture to machine a certain part, a sub-assembly re-design (or drawing required), or a stress calculation for a heavily loaded item. You are sure something like this has been done before but you can't locate the relevant information. Your regular source of knowledge, 'Old George' the filing clerk, can't remember (or worse still he has retired and no one can find anything—and this is a regular occurrence!). So what do you do? The job needs to be done so you do it and duplicate the data involved, complete the task and file the data.

A similar situation may arise in a few months or years and digging into the files may actually reveal one or other, or indeed both, sets of data. If they actually turn out to be the same then there is no problem, but if they are different in any way how do you choose the correct one to use? Were there any problems with one which were corrected in the other? How do you know?

Another example of data generation can be seen from a distrust of data filing procedures and ownership of data. How many of us receive information which we consider to be useful or important and instead of noting where we can get hold of it when required, we copy it and file it ourselves? Take perhaps a 5/6 page market/sales report which is circulated around a department of 10 people. If each person has a similar thought process then we waste 60 pages of paper, a lot of time, resource and energy copying it and unnecessary space storing it. If we had confidence in the department's filing systems and the integrity of the people who used it, we could allow it to be filed centrally and accessed as and when required. Suitable check-in and out mechanisms could be established to inform someone looking for it if it was already out and who that person was. But do we use these types of system? Well very often we don't, because sometimes the systems are difficult to set up and maintain—the very role that EDM seeks to address. A correctly installed EDM system should be able to provide these types of functions for a wide range of 'documents' of different types (paper, digital, microfilm, video, etc.). It should be able to provide the security required and to check data in and out to avoid the situation of losing data while access is being provided to someone else. It should allow read-only and update access and revision control procedures to ensure everyone is using the correct revision when documents are accessed. An EDM system should also provide full configuration management (CM) facilities to enable the correct chaining of documents up and down a structure. Figures 2.7 and 2.9 show typical examples of product histories and configuration records.

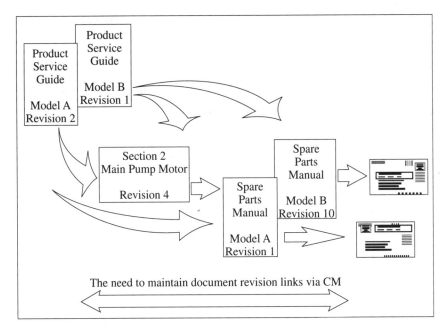

Fig. 2.7   Document configuration management

Many readers will relate to these kinds of problems in one way or another. Others may say, 'So what? My job doesn't require this kind of concern regarding information.' Well maybe not, but in almost all cases someone somewhere uses data either initiated or fully generated by someone else and EDM helps us to control data over the entire relevant sequence rather than on an individual basis as we may have done in the past.

In summary, EDM is a database tool to provide greater control over the data relating to individual products and one of its major roles is to assist in enabling a concurrent engineering environment to exist where previously there was none. In this way the full benefits of CE can be achieved with the appropriate level of data to back it up and assist in the drive towards a full CIM environment leading to business excellence.

We have already described EDM as a 'unifying' technology—other systems (manual or computer-based) must exist first for EDM to bind them together. However, it is also an 'enabling' technology—it enables business benefits to be realized such as:

- It acts as a conduit for the storage and communication of ideas, changes and authorizations in a manufacturing enterprise. No longer are different systems used depending on the data or systems involved in the generation of the data. A common approach to the storage and dissemination of data can be utilized, leading to the increased acceptance and confidence of a solidly implemented solution—'trustworthy' data rather than simply 'accurate' data.
- It enables all necessary people to work with the correct versions of documents and data—especially important as the concept of teamwork becomes more popular.
- It acts as a fundamental part of an integrated computing strategy—CIM cannot exist properly until the management of the associated engineering data is tackled.
- It provides quick and easy access to product data—an aspect of increasing importance to many companies, especially those in fast-moving environments and aerospace or defence industries where heavy reliance is placed on the associated data.
- It improves the use of standard parts and assemblies used throughout an engineering enterprise. Where, for example, a range of roller bearings is used in an end product, there may be a wide scope for the designer to select a range of sizes from a number of suppliers thus involving different configurations of end product and related information and drawings. With EDM the designer can be channelled into using a more 'standardized' range of such products with subsequent direct savings on procurement and stock-holding costs.

- It acts in accordance with and facilitates the introduction of quality standards such as ISO 9000.
- It provides facilities to enable a more complete product structure management. Very often there is no clear definition as to the exact nature and structure of the product's bill of materials or product structure and the different 'views' of it which can exist—for example as designed, as manufactured, as costed, etc. By working together on a common database each traditionally separate department can gain a better understanding of the product structure, leading to an early evaluation of the associated manufacturing processes.
- It provides a real-time data tracking and status information system for management.
- Its role is to provide a firm IT base from which other technologies can feed and grow.
- It can be used to focus a company's strategic aims in the specific areas of document management, configuration control, change management processes and so on. Figure 2.8 shows this concept. The 2D slices attempt to show the overall processes involved in a particular discipline or area. The marked area within each of these indicates the particular subset of the discipline (e.g. DO operations and procedures) used by the individual company concerned. The arrows indicate the overall processes (e.g. configuration management) as used throughout the entire company.

In this way we can visualize the use of EDM within the confines of the DO (as marked) and each other area on an individual basis. We can also see how each process can be extended through all the company's operations to form a strategic EDM implementation—for example to use configuration management throughout the company rather than concentrate on a complete DO-centred EDM solution initially.

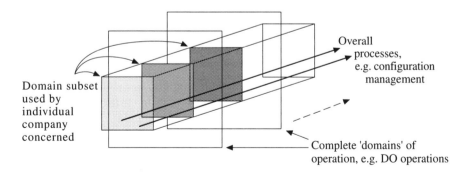

Fig. 2.8 The views of EDM within an organization

2.5   Key elements and functionality

As we have stated in previous sections of the book, there are no hard and fast rules relating to EDM systems—each offering in the marketplace differs in some respect, for example the functionality or user interface provided. Indeed, this is one of the main areas of contention in choosing a system: due to the highly 'personalized' nature of many companies' EDM requirements, many early EDM pioneers (and some current-day implementors) have adopted the approach of writing their own solution from scratch using a relational database management system (RDBMS). Others purchase a more readily available solution, but the aspect of tailoring the package to suit individual requirements is very important and this topic, together with others involved in selecting a suitable system, will be discussed further in Chapter 7. Since no clear definition of an EDM system exists it is hard to define exact functionality but the following are now becoming generally accepted as the standard functions which an EDM system should have, or be capable of supporting.

An underlying database

A modern, structured EDM solution requires some form of database management system (DBMS), which is a computer tool used to store the data concerned and provide a mechanism for manipulating and controlling it.

Extensive facilities are normally provided for the input, storage, updating, retrieval and reporting of data with a definition or 'schema' to define data formats, sizes and relationships. A structure or query language is also made available for enquiry and reporting purposes. In general each part, sub-assembly or main assembly will be held as a discrete entry or series of entries in the database. The 'fields' or attributes of each entry must be capable of being customized to suit particular requirements, for example part number structure and format, description fields, owner, remarks, quantity information, relevant project, revision level, etc. This is often termed 'metadata'. Data from other sources should also be able to be referenced here (via automated procedures linked to the relevant application program) such as electronically scanned data input, paper drawing file references, 2D CAD data files, 3D model files, NC tapes, CAPP data, etc. This is represented in a logical manner by Fig. 2.9. Most EDM systems use a relational DBMS (such as Ingres, Informix, Oracle or Sybase) in preference to a network or hierarchical model due to their ease of implementation, use of the Structured Query Language (SQL), processing capabilities, etc. Database types, structures and associated tools are discussed further in a wider context in Chapter 4.

The term design release management (DRM) is often used to describe the storage of EDM system controlled documents. These documents, datasets or files are stored in areas controlled by the EDM system, but the EDM system itself does not generally know about the internal structure of them. The

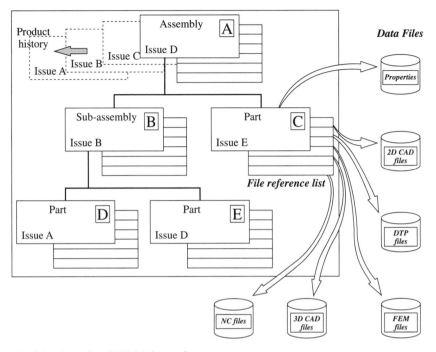

Fig. 2.9 Associated EDM data references

storage required by the metadata is small compared with the storage required for the associated documents—typically this may range in the order of 10 Mb for the metadata to 10 Gb for the files or documents.

Metadata can be generated within the EDM system itself, but can also be created by the attached applications (e.g. CAD/CAM) or can be taken from paper-based documents. Regardless of where and by what means the data is generated, EDM systems provide all the information about the data, including what, why, when and by whom it was generated or manipulated.

A data structure

The RDBMS structure employed with an EDM system should also have the ability to maintain product structure relationships and historical data (i.e. issue history) for engineering change and traceability purposes. The relationships are usually held via a product structure, or bill of materials (BOM) file to enable single-level parts lists, full multilevel explosions or where-used searches to be performed. The BOM/PS data should be able to link upstream to CAD parts list systems and downstream to MRPII BOM files, or utilize the same files where possible to reduce transcription errors and redundant data sources.

Historical information should be maintained by the system on completion of a suitable engineering change control procedure to enable past issues to be held to comply with the traceability requirements of current standards such as ISO 9000. In the past the representation of these types of product structure proved extremely difficult but the majority of current EDM systems now utilize a graphical representation format, some similar to Microsoft Windows or OSF/Motif and some of a more traditional engineering-type layout. This approach gives a better indication of the overall structure of a product than a purely textual based system.

Access control

Various access controls are required on an EDM system since much of the data originates from, and impacts on, many different departments, and therefore the consequences of error or unauthorized access are increased. Access is usually considered in a tiered manner:

1 Computer system access and top-level menu options. Very often systems can be configured so that after user login, control is immediately passed to the EDM system. The EDM system can then take the necessary action to check whether the user has authority to proceed further.
2 Medium-level access to documents/records via user status. For example system manager, draughtsman, industrial engineer, section leader, etc. Each one of these classes of user may then be presented with a particular menu of operations to choose from, some of which may invoke the application software required to modify the data type under review (e.g. CAD or DTP)—the user may not even be aware of some of the applications that are available on the systems to other users with legitimate access. This may seem restrictive, but is an important defence mechanism against unwanted intruders, and safeguards the confidentiality of sensitive information.
3 Low-level access to database information, perhaps restricting sensitive information such as costs or development data. This is usually at the lowest level of EDM system access and is closely aligned to the basic computer file access options such as:
   - No access
   - Read-only (ability to view or print data)
   - Read-write (ability to copy and/or amend data)
   - Read-write-delete (full ownership rights)

Document status flags

In order to be used successfully, EDM systems have to replicate the way engineers and manufacturing industry operate—the computer 'mystique' must be hidden. Engineering documents usually undergo a series of status changes as they proceed through various company departments, and tools to enable these status changes have to be provided in addition to the above access controls which are very often 'hidden' from the normal user. The generally accepted computer-based concepts such as amend, copy or delete are usually replaced within an EDM system by a number of status flags for each relevant document or file. An example series of document status flags is shown in Fig. 2.10. These are often used as part of a formal change control process.

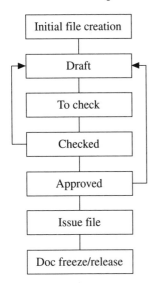

Fig. 2.10   Example document status flags

The exact nature of these will, of course, depend on the needs of the organization concerned and the individual EDM system used, but the following provides a brief description of those shown in Fig. 2.10:

- *Draft*   Seniority within an organization should not necessarily guarantee full access options to all data with a 'draft' status. Perhaps data viewing should be allowed but it may be in the company's best interests to leave the editing or updating to the person who created the data or who is defined as the current owner. Read, write and delete access options are automatically set for this person to enable the required alterations to be freely made during the design phase. Non-owners of the data may be given read-only access here, possibly on a project basis or only for a particular range of parts.
- *To check*   The owner of a 'draft' status document has the right to change its status to 'check' to reflect the transfer of responsibility within the data's life-cycle, and this is usually recorded as a system transaction with a time and date 'stamp'. During this phase it is probably sensible that the data

owner's access rights are reduced to read-only in line with the person authorized to perform the check. However, the individual organization concerned may decide that the file/data checker should have read/write/delete access to enable corrections to be made directly.

- *Checked*   The person authorized to check a particular file (and only that person) should have the ability to retract the status back to draft if the check completely failed or to advance the status on to 'checked' if it has passed. Again this is a 'sign-off' action and should be logged as such (i.e. a system transaction). If this procedure is followed at each transaction then it will be possible to establish a full historical record of each status change, who performed it, and the result.

- *Approved*   The file or data will at this point require to proceed to a stage where approval is granted for its general use. It will generally wait in a queue to be approved (or otherwise) by a person with the appropriate authority. While awaiting approval, the owner, checker and others should not be able to alter the data, so read-only access would normally be set. Again, if successfully signed-off the status would be changed to 'issued' (or perhaps directly to 'released'), if rejected the status may again revert to 'draft' to enable the cycle to be repeated.

- *Issued*   This step in the cycle may or may not be included depending on the system or the preferred method of working selected by the company operating the EDM system. For example, the file concerned may be issued for general use depending on a number of factors such as a drawing whose issue and release depends on others related to the same project or assembly; or the issue of a 3D model may rely on the results of an FEA stage being successfully completed.

- *Release*   At this stage the document would be given a release or issue number and made available for general use according to the rules established— perhaps read access with optional copy, plot/print capability—but the file should remain protected against all unauthorized edits. It is not normal to allow a file or document, once in this state, to be moved back to an earlier status. Where the document requires to be up-issued, a formal change procedure should be used with a copy of the released file being made with 'draft' status. It should be allowed to proceed through the stages we have outlined and, once issued and released, the original copy should be given a 'superseded' or 'obsolete' status flag.

As can be seen, procedures such as these (which are now known under the generic term 'workflow') can be time consuming and somewhat regimented but are a formalization of what should already be happening in a correctly structured environment. If proper checking and approval procedures are adopted then users can use data with increased confidence in the knowledge that the latest version is being used and that the master copy is securely held.

We can also see that someone has to administer the access controls provided by the EDM system—a point often overlooked when systems are being selected and installed. We will consider this aspect further in Section 8.2.

Data searches

Many figures have been issued to illustrate how much time engineers spend in routine tasks such as information retrieval—these range from 15 to 40 per cent. Whatever the real figure is it involves tying-up a trained engineer in a time-consuming clerical task. One of the main functions of an EDM system is to provide the search-type functionality required for today's engineering environments. Much engineering design involves new concepts and components, but equally much of it involves the re-use or adaptation of what already exists. In such circumstances, designers and engineers require ready access to as much data as possible to enable them to investigate other similar designs, products, sub-assemblies and parts (what worked and what didn't, who was involved and what were the related projects, for example). It may be that a new sub-assembly is not required, as originally thought, but an original design can be re-used with a few engineering changes where necessary. As we have discussed, the downstream benefits of such a scenario can be enormous. The ability to search for relevant data in a regular or *ad hoc* manner is therefore required and routines provided to format the output easily both on-screen and via report formats. If these routines are not made available, the inadequate feedback from earlier work will probably impact on the quality of any new designs. Old mistakes which could have been avoided will be repeated unnecessarily. Examples of routine reports or enquiries are:

- Standard drawing register listing
- Listing of each department's ECRs due for review
- Project document status listings

Examples of *ad hoc* reports and enquiries are numerous and varied and, depending on the flexibility and user-friendliness of the EDM system, are really only limited by the user's imagination. Examples include:

- Who last modified a particular document?
- List all major status changes to project A over the last two months.
- Print a list of ECRs associated with a particularly troublesome product.

Change management

The management of engineering change, ensuring that the latest version of modifications, drawings and specifications are in the right place at the right time is one of the most critical tasks facing engineering management today. This process frequently determines as much as 70 to 80 per cent of a product's final costs. Although engineering changes will never be completely eliminated it is important that they are minimized, especially in the early stages of product design. Design teams generally believe they know what their customers want (who needs marketing?) and design, detail and prototype, only to find out that they hadn't got it quite right. The net result is generally a flood of late engineering changes which simply waste money. Hence the need to get it right first time. An EDM system should be capable of raising and tracking Engineering Change Requests (ECRs), raising and circulating Engineering Change Notes (ECNs) and interfacing the change effectivity via date or batch/serial number onto an MRPII system for subsequent use. Once again, appropriate audit trails must be kept.

An ECR should be the first step taken once the need for a change is perceived. In general, this document will explain the details of the change, the associated reasons and likely ramifications, and should be circulated to all relevant parties for their comment and approval to proceed (or rejected with suitable explanation). Here again, the relationship information which can be stored with the data documents in an EDM system can prove very useful by providing a ready source of information regarding the 'knock-on' effects of a change to one part—what others are affected and what are the associated consequences. It is always better to have prior knowledge of these before recommending a particular change.

Assuming the ECR has been approved by all relevant persons, the next logical step is to raise and communicate an ECN, which is a document to notify all users that the ECR has been successful together with an indication of the work involved in the change and its likely effectivity date or revision. If, for example, the ECR required that a drawing be changed, the latest released version would be copied into a suitable 'workspace' (to allow ongoing access to the existing version) and the change code and status modified accordingly. The procedures described under the section on document status flags would then be followed. By utilizing such a change control process, the number of unauthorized and *ad hoc* changes to a product can be controlled where perhaps previously there was no control. This control can then be channelled into improving the process and the subsequent cost of the entire process. An example change control process is shown in Fig. 2.11 and reflects a complex change control process to cater for the minority of highly complex and risky changes together with 'fast track' options to allow the optimal processing of 'run of the mill' changes of simple to medium complexity and risk.

Configuration management

The aspects of traditional configuration management (i.e. the management and control of processes and physical items through records, documentation and data) and data management (which addresses the overall aspects of the data involved) are now being combined into one acronym, known as CMII

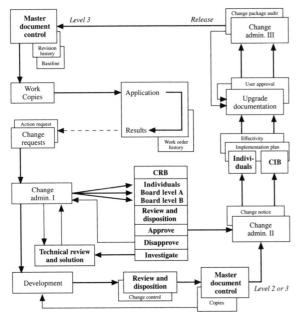

Fig. 2.11   Example change control process (*courtesy of the Institute of Configuration Management*)

(Configuration Management II). So while CM in its normal sense is a process for managing information about hardware, software and/or facility configurations and their associated changes, CMII takes this concept and expands it higher and wider, in a similar manner to MRP and MRPII. CMII and EDM share the same themes and are closely allied to:

- Total Quality Management (TQM)
- Continuous process improvement
- Concurrent/simultaneous engineering
- Total system integration

These concepts are shown diagrammatically in Fig. 2.12.

Configuration management II is defined as CM plus continuous improvement in the ability to 'change faster' and 'document better'. The ICM go on to quote the benefits as continuous improvement in the product or service, continuous reduction in cost, and continuous reduction in time-to-market. An effective CMII process also achieves all formal regulated requirements for configuration, records and/or data management. CMII emphasizes achievement of the highest levels of integrity in all documentation, records and data while accommodating change. We can see therefore that it requires a tool such as EDM to enable this goal to be achieved (Appendix A gives further information about CMII).

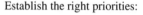

Four major performance improvement
initiatives are byproducts of CMII

Establish the right priorities:

    (1) CMII
    (2) Physical processes
    (3) Physical items

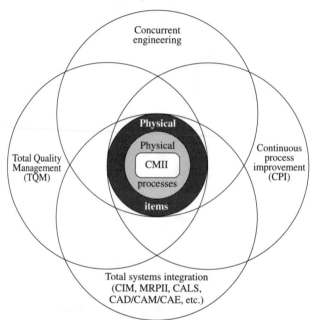

Fig. 2.12   The CMII process (*courtesy of the Institute of Configuration Management*)

An EDM system must be able to provide full traceability (or change history) over the entire product structure to enable information relating to the serial number or lot/batch of a product or individual component to be retrieved at any time. Again audit trails are important to be able to retrieve historical information. A representative change history or configuration record is shown in Fig. 2.13 (see also Figs 2.7 and 2.9).

Project management
The usefulness of an EDM system can be enhanced considerably if the data 'documents' are directly linked to projects, i.e. individual 'chunks' of effort related to a particular customer or product which need to be accounted for separately. One possibility could be to interface the EDM system to an exter-

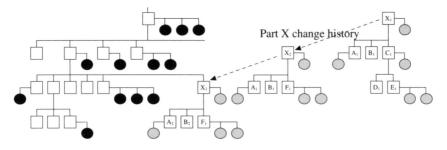

Fig. 2.13   Product change history.   □ Part, ● document

nal project management system. This approach provides a more comprehensive overall solution, but the benefits of a single, integrated solution which we have already outlined would not be so readily achieved. The best solution, assuming adequate functionality is provided, is to incorporate project management functionality directly into the EDM system, perhaps as an optional module. For example, functions such as project task initiation and subsequent tracking, man hours and cost reporting of actuals against estimates need to be made available. This is where an overlap of EDM and existing technologies can occur. One area has already been mentioned: the use of a common BOM structure for EDM and MRPII. Project management is another. A further option is to consider using the project management modules now being provided with a number of leading-edge MRPII systems, which perhaps could be interfaced with suitable EDM systems. Even they, however, require extensions to cater for the extra data type references previously referred to. A typical product/project hierarchy is shown in Fig. 2.14. By providing project management functionality as part of an overall EDM solution, a further layer of reporting and control can be established over what is becoming a more widely used and correspondingly important area of everyday engineering. Whatever method is chosen, the integration of EDM and project management provides clear business benefits and may well become standard functionality in future system offerings.

Message handling
Most EDM systems provide the ability to flag actions, warnings and messages via a simple action 'flag' system of initiating standard changes to certain database fields, or via their own or the host computer's electronic mail (E-mail) system. These messages can be used to determine and track the current system's status and should allow engineers and management to receive and respond interactively to these messages. This enables improved communications throughout an organization, especially when the individuals concerned are located in different departments, manufacturing plants or even different

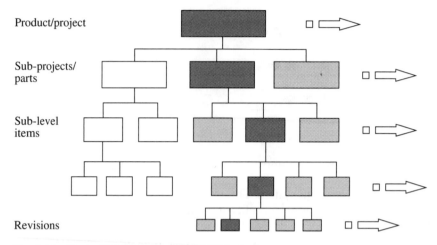

Product/project

Sub-projects/
parts

Sub-level
items

Revisions

Fig. 2.14 Typical EDM product/project hierarchy

countries, as is now becoming more common in considering pan-European and worldwide product development initiatives. Example types include 'Alerts' (i.e. a predefined series of event-driven trigger mechanisms), for example, 'Drawing XYZ is now ready for checking', and 'Messages' (i.e. a more general *ad hoc* message system) which could be used to store more verbose information, perhaps in association with an Alert. For example, 'Drawing now checked but rejected because of .... and I believe we should consider the effect of ....'. It is also important that the message system, if provided with the EDM system, is correctly set up and administered to ensure that each user understands what the messages mean and their obligations in receiving and acting on them, the means of enforcement and any relevant circumstances where formal acknowledgement is mandatory, etc.

File handling, storage and archival
EDM systems, because of their very nature, store and accumulate large volumes of data from a number of sources. They therefore have to provide:

- The ability to capture and manage documents by storing indexes to data in a distributed fashion over a computer network. Most EDM systems do not actually store all of the data files for every application on the computer system (these are generally held in some secure area, often termed an 'electronic vault', on the disk close to the application software). The indexes to this data, and the transaction records and control information (i.e. the metadata), are actually what the EDM system contains. It is therefore important that the EDM system can maintain the correct location of every file over a distributed computer network, and in many cases control access to this data 'vault' via various means (perhaps by holding local copies for speed of access but always ensuring that the integrity of the data is maintained).

- The ability to avoid data corruption during update operations. This is a very important aspect in a comprehensive EDM system—a well-structured and robust system can be of great benefit to an organization, but a poorly implemented and functionally deficient system can bring a company to its knees. We have already discussed how the document update process generally involves working on a master copy and replacing this in the 'vault' once successfully updated. Control of this, and other related file operations, especially over a distributed computing environment when perhaps a system fault is encountered during such an update process, must be strictly controlled and administered.

- The ability to easily archive and restore sets of data via specialist EDM routines (independent from the host computer system's data backup facilities). As systems are implemented and continue to grow, the data volumes and associated information correspondingly increase. More data files will be copied routinely onto permanent storage media such as magnetic tape, hard and optical disks. It is therefore imperative that appropriate backup and restore facilities are available. When performed via the EDM system, such facilities ensure that adequate records are kept of when backups were done and the tape or disk identification labels, etc. Therefore, from a user perspective, the procedures are simplified, more controlled and with the opportunity for errors considerably reduced.

Housekeeping and transaction reporting
Throughout the course of our discussions on EDM functionality thus far, we have mentioned many times that it is vitally important that a full record is kept of every transaction that is performed on the database. This is required for two main reasons:

- First, to act as an audit trail to track the various activities that have been performed against a specific document or series of data objects. In this way we could for example establish who had accessed a document over the past couple of weeks and confirm what operations had been performed on it (and indeed whether authorization was correctly administered for these accesses).

- Secondly, we could utilize the transaction file in the event of a problem occurring on the system which would require a data restore operation to be performed. This would generally be used in conjunction with a backup of the database from the last archive tape. The system could be restored from tape and the transaction file 'replayed' into the system to re-create the changes from the time the backup was taken to the time the system crash occurred. This is normally termed 'roll forward' recovery.

Many 'housekeeping' tasks are usually made available with the vast majority of EDM solutions. The most important of these have already been mentioned—that of document or data archival and restoring, and the need to provide facilities to ensure the correct system backup and restoring in the event of a system problem or 'crash'. The latter point is worthy of note since, to a growing number of large organizations, the consequences of a lengthy down-time as a result of a problem can be of a very serious nature. These companies are now looking for EDM systems which either incorporate directly, or utilize an underlying RDBMS system's, resilience mechanisms to minimize this aspect of database operations (via fast recovery mechanisms, on-line and incremental archiving and disk mirroring) to ensure the maximum on-line availability of the system. Other housekeeping facilities include tools to ease the definition of new system users, setting-up authorization files and sign-off signatories, document type definitions and server 'vault' locations, etc.

Many of the EDM solution suppliers listed in Appendix A will have individual 'bells and whistles' added to their products to provide unique selling points but the above functions constitute 'core' EDM functionality of the majority of good quality, currently available product offerings.

2.6 Relevant industries and application areas
Having considered the subject of EDM in some detail, we should now ask the question 'Where can the technique be used?'

In order to help answer this question, I would suggest we require to pose another: 'In which business operations do structured sets of information and data cease to be of importance?' Answer: none.

It is simplistic to assume that all areas of business can benefit from EDM, but it does highlight the fact that, in theory at least, there are no barriers to its use. We can see, by referring back to Fig. 1.1, the main functional areas within a manufacturing enterprise which can make use of EDM technology. There are no hard and fast rules here—these are perhaps considered as being the more accepted areas, but there is no reason why, say, a sales order processing environment should not make use of the configuration management information held in EDM if this would benefit the company concerned.

So the potential application areas within a manufacturing company are numerous. What about the overall market sectors where it can be used? Once again we need to consider the 'big picture'. The vast majority of companies want an integrated and seamless flow of information through their business, but what they actually have is a sequential flow of information, isolated islands of automation, an orientation around paper and drawings, low visibility of the actual design and development process, and a culture which resists change.

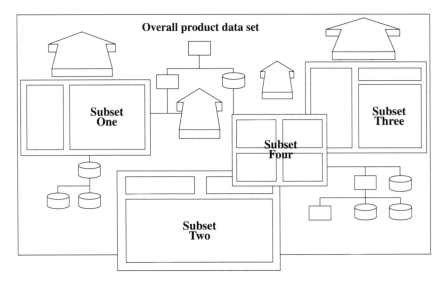

Fig. 2.15   Different views of an overall data set

EDM seeks to challenge this type of environment. Everyone uses and generates sets of related data and as such each data source will have a different view
of it, for example computer scientists, analysts, suppliers, engineers, field service staff. The majority of these sources claim to have originated technologies
to manage the data, e.g. PIM, document management, image processing and
control systems, database management systems—all have relevant but different views of the overall data set (Fig. 2.15).

In such cases the acronym used to describe the relevant tool, such as EDM,
can prove to be disadvantageous to its use. We have already discussed in
Chapter 1 how EDM can be misconstrued: 'I'm not an engineer so it doesn't
apply to me.' Many use the term EDM since engineering products today
encompass many data sources and technical systems: MCAD, CNC, ECAD,
MRP, FEA software, via PROMs and disks, logic devices, sensors, etc. But
equally the term PDM (Product Data Management) is also very descriptive,
and some would argue more relevant. For example, a software company may
need to use EDM for source logging, configuration management and change
control but would dismiss EDM ('We're not engineers') and would use the
term PDM ('We generate a product').

Similarly, someone may argue that PDM is not a useful technique for their
environment since they do not make a product. Perhaps they are a service-
based industry with a need to configure customer support records and configuration management information in terms of revisions and maintenance
schedules—Service Data Management (SDM)! And so it continues.

The beauty of EDM, despite the exact meaning of the TLA, is that once the

Table 2.1 Examples of use of EDM

Industry sector	Application area
Aerospace	Configuration management very important Documentation procedures must be adhered to Strict standards control
Defence industries	Similar environment Wider range of standards
General engineering	Traditional process, high-volume, batch, and job/project environments
Computer manufacture	Fast-moving environment Time to market critical Fast change management cycle
High-tech equipment manufacture	Similar environment
Computer software	Specialist tools are available here but EDM could prove suitable for a small environment
Consumer goods	Fast-moving environment CM and change management important
Service industries	Possible use here especially in change management
Oil and gas industry	Safety-critical environment Very high levels of documentation Project-related environment

concept is understood, it can be applied in many areas. In the end it is the same thing: the push for it may have come from the need to engineer an increasing number of high-technology products, but it is relevant to any enterprise that has a need to control a reasonable amount of data (in any form) related to the product or service they provide. For example, some examples of industrial sectors and associated application areas within which EDM could be used are listed in Table 2.1.

3

The benefits of using EDM

3.1 Project investment and returns

Obtaining an asset and not looking after it correctly can cause long-term problems and prove very expensive. Like most things in life this proves to be true in large- and small-scale situations. We have already seen how this lack of investment over a prolonged period has caused serious problems in traditional industrialized societies. A rather unusual example can be used to highlight this situation: the sewers in many parts of the UK, and London in particular.

These sewers were built during Victorian times and onwards to cater for a city which was probably only about one-tenth of its current size. Once built, they were very difficult to maintain and so have been the subject of minor maintenance only, while being subjected to levels of use far beyond their original design criteria. It is a testament to their original construction that they have survived so long. However, the situation has now got to the stage where they can no longer cope and are in severe danger of collapse in certain places. The capital investment required to restore them to the appropriate levels required for the 1990s is now so high that it is almost impossible even to consider undertaking the work—a clearly undesirable state of affairs!

So, instead of facing up to an ongoing cost of adequate maintenance and upgrading over the years, a serious problem has now manifested itself costing many times the ongoing maintenance costs. Actually doing something about such a problem at the time is not easy either. Consider the reaction if you were in Government during the period where adequate maintenance costs could have been allocated. 'I want to spend £Xm to maintain the service,' you say. Immediately everyone is up in arms protesting: your own party thinks that costs should be cut, not increased; putting up taxes to pay for the work will lose you the next election; the public are furious ('We need new roads and subsidies, not better sewers'); and the opposition attacks this 'squandering' of taxpayers' money. So the problem is swept under the carpet for someone else to consider. It is very difficult to stick your head up to take a long-term view of a situation when an army of opposition is out there ready to shoot it off.

The result: 'Short termism'.

No doubt you can think of examples within your own experience which serve to highlight the same issues. For example, it is usually easier to invest in a new CNC machine tool than a centralized NC tape preparation system to program them: 'Each CNC machine now comes with its own separate input mechanism included in the price, so why buy an external system to standardize?' It is usually easier to purchase some CAD terminals to improve time to market than introduce a new MRPII system which could potentially reap higher long-term rewards (less cultural impact, fewer operational problems to overcome, etc.).

EDM systems fall squarely into this category. Investing in a 'blanket' technology such as EDM is more difficult to visualize and requires a head-up, long-term view to be taken. The attitude of, 'We've got a CAD system and a ...., why do we need something else?' needs to be replaced by an informed and justified view of, 'If we invest X today we will achieve Y tomorrow.'

There is also the question of management scepticism to be overcome. 'I've seen all this before—it'll never catch on' or 'It's just another bandwagon to jump on' are statements which require to be correctly addressed at the outset. This can be achieved by correct education and knowledge regarding EDM, backed up by a suitable justification and action plan. Project investment in this area requires a different view to be taken—in line with business excellence.

It is not just the technologies themselves which suffer from 'short termism'. A recent study in the UK identified that 85 per cent of UK companies face problems in implementing concurrent engineering because of the short-term attitudes of top management.

However, all is not doom and gloom. The 1993 Manufacturing Attitudes Survey carried out by Computervision does show that long-term planning is increasing, belying (to some extent) the criticism of short-termism. Over 90 per cent of UK firms have long-term business plans, although these are mainly geared towards a 3 year planning cycle. Conversely, US firms in the UK look further ahead—71 per cent of them work to a 5 year plan. Just 51 per cent of UK-owned and 47 per cent of European-owned manufacturers do. This is shown in Fig. 3.1.

Investing in an EDM project, as with any major capital investment, causes certain fundamental questions to be asked. One of the main ones which relates to the industrialized West, and the UK in particular, is the oft-encountered accounting/financial stranglehold exhibited over many companies by management. It would appear that the main route to senior management, general management or managing director status of many engineering and manufacturing concerns comes primarily via the accountancy profession. Even in those companies where this is not the case the managing director generally places greater emphasis on the financial director's viewpoint as to whether to invest in the technology in question. No-one doubts the importance of the

Percentage of sites

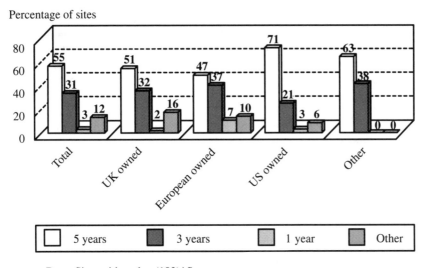

Base: Sites with a plan (183)/ Spontaneous

Fig. 3.1   Long-term business planning cycles (*source: Computervision*)

financial implications in any investment initiative—they must be included—but very often they override other equally important, and very often more important, aspects of long-term investment such as quality of product or service and engineering related benefits.

This is typified by a statement from one senior executive (presumably a financial director) at a recent seminar on investing in new technologies such as EDM: 'This is all very well, but my board won't sanction this unless the payback period is less than three years.' This view, while being perfectly acceptable fifteen to twenty years ago, must change. One is left wondering how long that particular company will remain in business before gradually withering and crumbling to a premature demise like a London sewer. Concerns such as these are now being expressed by a number of different business communities, not just engineering and manufacturing. For example, a quote from an eminent business professional in 1989 summarizes the situation well: 'Our fixation with financial measures leads us to down-play or ignore less tangible non-financial measures such as product quality, customer satisfaction, factory flexibility, the time it takes to launch a new product, and the accumulation of skills by labour over time. Yet these are increasingly the real drivers of corporate success over the middle to long term.'

We must strive to look past the traditional payback period and this predominance on pure financial performance, towards more 'open' techniques such as discounted cash flows and long-term project investment scenarios in order to actively invest in technologies which assist in enabling a concurrent engineering environment which in turn will lead to improved business excellence. Such attitudes also extend to the very top of the corporate ladder. The Computervision Manufacturing Attitudes Survey clearly shows that the manufacturing/City communications gap is alive and well (Fig. 3.2.). Despite lack of investment being identified as a major retardant to recovery, 41 per cent of listed companies and 84 per cent of unlisted firms do not communicate with the City. Little wonder that the City does not understand manufacturing if all it sees from companies is the annual report. Manufacturing must take steps to close this information gap if it is to stand a chance of convincing financiers of the appropriateness of investing, otherwise they will be starved of the financing needed for growth once the opportunity appears.

Justifying any form of advanced manufacturing technology in a more 'open' manner is not easy, since the returns on investment can be very difficult to define and may be highly intangible in their nature. For example, one recent UK study quoted that companies can save up to 10 per cent of turnover by adopting an enterprise-wide EDM solution. While this is no doubt possible, such a statement is too vague and generalized to be put forward as a suitable justification. Very often the decision to invest in such technologies comes down to answering the question, 'Do we want to remain in business or not?' and in many cases relies on the faith, trust and vision of the future held by senior management to steer an organization in the right direction.

Once on the correct course, however, the returns from investing in newer

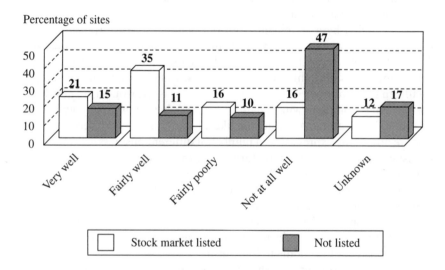

Percentage of sites

Base: All sites (201) / Prompted

Fig. 3.2 How well the City is perceived to understand what companies are doing (*source: Computervision*)

technologies such as EDM can come in many forms and are usually split, as we shall see later in this chapter, into large-scale, mid-level and a number of intangible factors.

3.2 The history of large computer project investment

Many readers will have experienced some form of major computer system investment. Indeed, many may have experienced a number of such implementations, each promising to address the relevant business deficiencies or perhaps address the shortcomings of the previous system! This could be the implementation of a computerized employee database and associated payroll system, a manufacturing control or MRPII system or perhaps a major investment in CAD. Whatever the system may have been, the timing associated with its installation and implementation will have played a vital part in how this function was accomplished. Many boards in engineering companies are now beginning to ask why many of these investments have been so disappointing, why IT is treated as a cost centre rather than a strategic competitive weapon, and so on. To consider such questions fully we have to look back at the history of computing in industry.

For example, the traditional approach to introducing some form of information technology in the mid-to-late 1970s was to hand the task over to someone else, namely the 'DP' (data processing) department of the company (everyone had a mainframe-based DP department!), so the supply of a system became 'their' problem. 'We just use the system, but they don't even ask us what we need or want.' System users had the luxury of blaming all the problems on the DP department and slating the system as being unusable—the 'DP wars' of this period became a serious problem. Such a centralized computing philosophy gained considerable status in the eyes of senior management, often forgetting its role in serving the business and justifying its existence by spending more and more money to meet ever-increasing development deadlines.

The scene seemed to change, however, especially in the engineering sectors, around the early 1980s when systems started to appear not only for use by engineers, but also for set-up, administration and control by the same engineers/users. 'Ownership' of IT started to appear with, for example, early CAD/CAM systems. No longer could users blame the DP department if the system did not perform: it was their own colleagues or perhaps even themselves who were to blame. This aspect of ownership (which incidentally did not go unnoticed by the DP managers!) was further extended in line with the increasing use of mini-computers throughout the early to mid-1980s, where the computer itself could be situated in a normal office environment, so the air-conditioned room of the DP department was no longer required. This decentralized computing trend has of course now further developed in line with the PC 'revolution', but at the time, engineers having control over their own computing resource was fairly revolutionary.

What is basically being said here is that, over time, current thinking on how best to implement and introduce IT solutions has changed, just as with other aspects of life where we are continually learning and refining our knowledge. Since it is a generally accepted principle that we use this past knowledge and experience to avoid us repeating the same mistakes over and over again, here we outline some previously used (and some currently used) approaches to the question of how best to introduce a new IT solution such as EDM. In this way we can better understand currently adopted methods and practices.

The 'jump-in' approach

This approach is typified by a technology-oriented management style where the view is taken that by throwing a new technology at the current problem it will immediately be solved. While it is still true that we need certain people to jump in and 'test the water', the current competitive climate reduces this number considerably due to the high risk factors involved. Very often this risk can be minimized by financial and support incentives from research and development establishments and Government bodies but there is still a considerable fear of being left using a technology that is now non-standard, expensive, cumbersome or plainly ridiculous.

This approach can also be used to demonstrate the 'moving bottleneck syndrome' to great effect. By not considering the problem from an overall perspective, the immediate problem may be solved but results in an even greater problem appearing elsewhere.

The 'don't be first' approach

This is the exact opposite of the previous point. This approach takes the conservative viewpoint of always waiting until someone else takes the risk involved in a new technology. In this way it is hoped that all the pitfalls will have become clear and a less expensive route will become apparent.

In some cases this may ultimately prove to be the most expensive. What happens if the new solution/system is a real winner and by taking the lead in its use a company could steal a two-year lead over its rivals? By being an 'also-ran' in waiting, no such profitable gains or savings will manifest themselves. In many cases the 'management of risk' requires to be carefully measured in choosing between this approach and the previous one—the justification for choosing one or the other is certain to change for every specific case.

Employ an external consultant

Traditionally, many managers have abdicated their responsibilities in deciding how to introduce a particular IT solution and decided to employ the services of an external consultant to take the control of the project. There is nothing wrong with this when put in perspective but very often this approach has been used due to a lack of technical awareness regarding the problem under consideration. Perhaps a director or member of the board would take the 'easy option' of subcontracting the problem to someone else because of a lack of knowledge of the problem itself or of knowing where else to source information and help.

In many cases the consultant is given the task of introducing and implementing the system with a 'call us when it's ready' attitude by the client—in effect acting in a similar manner to an internal DP department. This devolution of responsibility has been proved to fail in the vast majority of cases, for many reasons. The current trend is very much more oriented towards a true 'consultative/hand-holding' approach where help and assistance is given to ensure the client can take the initiative and responsibility for the introduction of the system, based on an improved knowledge and understanding of both it and its associated consequences.

An unstructured approach

This can be illustrated by the expression 'putting the cart before the horse'. An example of this could be purchasing a face-lifting tool (i.e. a software application which allows older systems' user interfaces to be remodelled to provide commonality and look more advanced) without an overall strategy being in place, or perhaps purchasing EDM before any other software solution. There is nothing wrong with this approach if the missing applications can be slotted in when available, but very often they cannot and an unstructured IT strategy prevails. A successful IT strategy requires a 'roadmap' of how to achieve your long-term goals with suitable 'on-ramps' for new technologies to be incorporated as required. Immediate problems do require to be addressed but should be able to be done with the confidence that, in the future, FEA or shopfloor data collection or any other solution can be integrated into the system. We have already seen that a business strategy is essential and it is important to have a vision of how IT can support this. In the past an approach such as this could suffer greatly because of the lack of adequate standards. The IT strategy will define standards which can be adhered to as the various problems that need to be solved are picked off. This situation has improved considerably since the introduction of an open systems philosophy and the corresponding increase in the use of corporate standards. This aspect of computing technology related to EDM is discussed further in Section 5.7.

The 'mega' approach

This is another example of a technology-based approach where rather than attempt to find a standards-based route through the integration of a number of different applications, it may be decided to retire the majority of the existing systems and replace them. This approach will become even less popular than it was due to the very high cost factor involved and the increasing number of standards available to assist in communicating between disparate systems.

One of the interesting factors in this approach, or indeed any approach which involves retiring a software application before it has served its anticipated lifespan, is the fact that no one likes to admit that the initial purchase of the software was a mistake. If you were responsible for the purchase and implementation of an MRPII system two years ago which cost your employer £150000 with an estimated life-span of at least five years and questions were now being asked about its suitability or future, you would hardly be likely to agree! 'Oh no, this is the best investment we've made—and I'll prove it' would be more like it. The result is often a far from ideal system being made to perform tasks or roles it was never designed for, with associated lack of functionality, responsiveness and an escalating implementation budget. It takes a brave person to admit they were wrong in such a scenario and to further commit themselves to a further global company computerization plan.

The hesitant approach

There is a common question posed by project managers and consultants: 'How do you eat an elephant?' Answer: 'One bite at a time.' This is very true and is always worth bearing in mind on any large project, be it IT-based or otherwise. However, it begs a further question: 'How large a bite do you take?' It is impossible to define an optimal size for every project because so many individual factors will apply. Many people who react against the previous approach very often go too far in the other direction and take every step of the project at too slow a pace. The need to be seen to prove each project milestone is generally appreciated but often a more aggressive timescale requires to be set in order to meet more strategic business deadlines. The setting of IT project implementation timescales of five years and over will be discussed later in Chapter 7. Suffice to say here that maintaining momentum and commitment over such a time-span is very difficult. For example, a large UK-based aircraft manufacturer launched a major initiative about eight years ago to integrate their engineering and manufacturing computer systems together over a five year period. After committing over £40m to the project they now admit its failure and are reviewing their future strategy in this area. But they are not unique here—a recent UK study showed that at least 60 per cent of all major IT projects were either twice over budget or 50 per cent behind the stated delivery date.

'So what's changed?' you may ask. Well, a number of significant advances have been made recently in our understanding of the areas of successful project management. Many of the traditional computer suppliers such as IBM, DEC, Bull and ICL have made significant moves into the consultancy and services market, supplying systems integration skills covering project and risk management, to assist in implementing more successful IT solutions.

Current thinking now stresses a number of levels of pilot implementation under a well-defined major project strategy with each step being documented and approved in a logical manner as required. The use and understanding of project milestones and of associated PC tools to address the aspects of project management are now widely accepted, as is the importance of continuance of investment throughout the life of the project. The importance of people is also now a more widely accepted aspect of the successful project management of information technology solutions such as EDM. This may sound obvious but the old saying, 'Computers make it possible but people make it happen' is nowhere more true than in an area such as this. The awareness, training and involvement of all relevant people is very important both during the investment justification process itself and in the subsequent implementation and use of the system. Many companies rely solely on specifying the required functionality without full regard for the people who will run and use the system. Also, very often after installation, the way people work with computer systems provides an excellent example of Parkinson's 'Law of Triviality' which states: 'People devote their attention to subjects in inverse proportion to their contribution to the business.' The computer (both the hardware and software) becomes that subject, and large amounts of time (and consequently money) are devoted to it with a disproportionate amount spent on the people who will make intelligent use of it. At the end of the day, the computer system is simply a tool to allow a number of persons to perform their jobs more effectively.

3.3 Suppliers' guarantees

There appears to be a train of thought in industry today that suggests that the installation of a computer system automatically guarantees the quoted benefits. This is clearly not true—EDM systems, like many other computer-based solutions, generally do not carry guarantees.

Shrink-wrapped software such as readily available word-processing or spreadsheet packages have a more generally accepted 'extent of supply', i.e. they provide a set of functional procedures which either work as intended or they do not. You can normally return to a high street store and complain if a product does not function as explained in the documentation. An explanation of where you are going wrong, a replacement or your money back are the normal courses of action here. The situation becomes more open to problems and misunderstandings as the software solution increases in complexity and size. Very large computer implementation projects require to be carefully managed if the anticipated benefits are to be realized and all parties agree a satisfactory conclusion to the investment.

Medium-sized projects, such as the procurement and implementation of an EDM or MRPII solution, are subject to similar problems, albeit of a lower magnitude. These problems can, however, prove to be major obstacles in the way of a smooth and ultimately successful implementation.

One of the major reasons for this is that the expectation levels of each party involved in the IT solution are incompatible. You, as an implementor of EDM, may expect it to perform a certain operation as standard (perhaps without proper qualification or proof). After the purchase of a system which ultimately does not perform the operation in question, you may well feel the implementation to be a failure because of this one point. Such a situation may result from both sides of the selling situation, from both client and supplier. Let us consider an actual example. Many of you may remember the glorious days when early MRP systems were going to solve all your manufacturing-related business problems. Bold claims were made about the levels of improvements which could be made, such as 'Reduce inventory by 40 per cent' and 'Guaranteed to improve on-time deliveries by 25 per cent'. In certain cases minimum benefit levels were quoted and many implementations were deemed as failures because these levels were never reached. So where does the fault lie in situations such as this?

While the saying 'Let the buyer beware' is as true today as it always was, suppliers who use false claims and those which cannot be adequately qualified can now find themselves subject to heavy law suits or claims if subsequently proved in court. Generally speaking this results in a more factual representation and explanation of a product's overall capabilities, at least at the higher end of most market sectors. In certain cases identified, specific objectives may be agreed prior to the purchase of a system as part of an initial scoping/consultancy exercise. In this case of course the benefits have been quantified and agreed, so if the exercise has been correctly handled, there should be no cause for subsequent disagreement.

Conversely, of course, not all the blame can be placed on the supplier in our example either, since the purchaser may not have taken adequate steps to qualify the statements being made and set goals relating to his particular environment. For example, the inventory reduction figure of 40 per cent may have been based on the supplier's previous customers who were focused on a high-volume, high-technology batch production environment. The supplier may never have meant to mislead you into believing this should also apply to your jobbing shop/low volume engineering plant.

Each party has the responsibility to understand and explain needs and requirements together with the ability and method of meeting them—a well-organized selling environment and one in which there are up-front, agreed deliverables is preferable.

Contention in an IT implementation, as with other project areas, can come from many sources, not just the functionality of the system itself. More often in today's environments, a teamworking approach is required to install a system—a partnership situation where an open and trusting environment is generated, can often make or break a system implementation. This is an important point which will be considered further later: when choosing a solution such as EDM, functional requirements must be met, but so too must the final implementation be deemed a success, and so the ability to work with the people concerned is very important. You may purchase the best EDM system on the market but still not have it implemented in five years' time for a number of reasons—perhaps you find it difficult to work with the supplier's consultants or their quality of implementation is poor. So, although the choice of software seemed correct at the time, the implementation is classed as a failure—what effect would such a situation have on your career progression?

One other point to consider if a project does start to go seriously wrong or fails completely, is that of 'finger pointing', of moving the blame onto another party. Guarantees may not be available but documented proof of agreements may have been made and the correct management of the project, and setting of original expectation levels, may ultimately result in the project returning to a more stable and agreeable state. Very often a three-way implementation team is involved with supplier, client and external consultant. If this is the case, it is important that the client recognizes and takes ownership of the problem and provides the required leadership to guide the other parties along the desired route. Too often it is assumed that the problem will be solved by the consultants alone and a further source of misunderstanding and possible acrimony is introduced. Each party involved in a project such as the introduction of an EDM system needs to understand that they have a role to play and what that role is. This should be defined and a suitable structure, responsibility matrix and project management team established to achieve the required goals. The parties involved will of course be case specific but from a client's perspective could typically include:

- *Supplier*   These could be of various types, from one or a number of sources. For example:
  - hardware supplier
  - software supplier
  - implementation and consultancy services
  - systems integration supplier
  - training supplier
- *Project manager*   Timescales and planning
- *Users*   Training requirements
- *Administrator*   System set-up, policy decisions and data take-on, etc.
- *Steering committee*   Project review and planning

The topics outlined thus far will be considered in more depth later. It is hoped that by briefly reviewing them here we can see that there are a considerable number of factors which, if successfully managed, can ultimately lead to a successful implementation for all concerned.

3.4   The strategic benefits

As with any item of a strategic nature there are a number of benefits which can be gained from its correct adoption and use. EDM has a number of such benefits. For example, one UK electronics company has reduced its overall design lead time by 50 per cent and its time to process changes from 14 to 4 weeks. Each installation will provide not only its own unique benefits, but will also contribute to the overall synergy of the entire process, of the combination of all the benefits considered together. These benefits cannot be ranked one above the other since each are very case-specific and indeed the classification of strategic, mid-level and lower-level benefit will also vary from enterprise to enterprise. Each organization's vision of the future and the relative importance of addressing the particular topics to reach this vision will change. What we can see, though, is the increasing level of benefits which can be achieved by considering EDM at the strategic level rather than simply as an electronic filing cabinet. This is shown in Fig. 3.3 and also relates to the types of benefits which we will consider in the next three sections of this chapter.

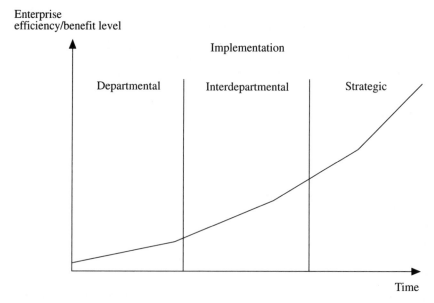

Fig. 3.3   EDM: benefit 'banding'

The following examples of strategic benefits are therefore not in any form of priority order and should not be considered exhaustive.

The first benefit which a company may consider under this heading is one which we have already touched on earlier: the fact that EDM can drive the move towards (and support) a concurrent engineering environment. It is very difficult to imagine a comprehensive, reactive, state-of-the-art CE environment existing within a modern high-technology manufacturing environment without some form of structured engineering-related database 'foundation' upon which to act. We have already seen how competitive advantage can be gained by adopting CE principles—by early involvement and evaluation of the manufacturing process further upstream in a product's life-cycle, by quick and easy access to product data, by more complete product structure management and by being confident in working with the correct data and revision information. These are aspects in competitively trading in a modern environment for many engineering and service-related companies which, if not already achieved, are being actively pursued, in certain cases merely to remain competitive rather than make a quantum leap forward over the competition.

The second benefit in considering EDM is that, because it is a technology which can involve many other related areas, it tends to give a more solid backbone to a long-term IT strategy—it provides the structure to consider the adoption of these other technologies where perhaps previously there was no means to do so. The adoption of a policy regarding RDBMS systems is something which a company would probably consider without the catalyst of EDM, but perhaps general document management and scanning technology is something which may have been considered for some time but stayed at the bottom of the list because it was always 'leapfrogged' by a more important requirement. EDM may provide the structure for such plans to be put in place, and actually happen, as part of a much larger exercise than being previously considered on its own.

Many organizations do not have any form of coherent data management or records strategy, to define for example how information is produced, stored, retrieved and accessed. Such a strategy should consist of policies and procedures that indicate how records should be managed to maximize information benefits, while minimizing related costs. In the USA the concept of a Chief Information Officer (CIO) is readily accepted, to provide control and direction to both IT and non-IT information. In general, however, and in the UK in particular, data records are given low priority by organizations, which see records management as a non-strategic and somewhat menial task. This is exemplified by one recent assignment by PA Consulting Group, who were asked to help in the justification of an EDM solution for a large UK industrial organization. They found the annual cost of processing, maintaining and storing vast amounts of paper-based information added-up to a staggering £1.8m! Records were spread over many different departments and locations, each of which managed their own budgets. There was no single place where the total bills for documentation were consolidated and thus the management of the company had never seen this as a strategic issue. Characterizing the costs in this way gained management attention and moved the project forward.

However, it is not just the storage and maintenance of data which is important in building a data management strategy—there is also a legal risk. A recent UK survey found that 40 per cent of large companies either fear, or have already experienced, litigation due to lost or missing records. So, EDM can be used as the catalyst in a company to consider a company-wide data and records management strategy, particularly where the stored data may be required for a considerable time (100 years in the case of nuclear installations), and break out of the vicious circle of paper (as shown in Fig. 3.4) which, it is claimed, can account for up to four weeks of the average executive's time every year waiting for filed information.

EDM may also permit any existing networking and distributed data storage capability to function in a more effective manner. In addition to this system management aspect, EDM can also assist in the more effective use of the existing software applications by providing a uniform user interface or data entry mechanism, a common 'environment' to enable more familiar system entry paths to exist.

EDM technology can also be used to improve the image of an organization.

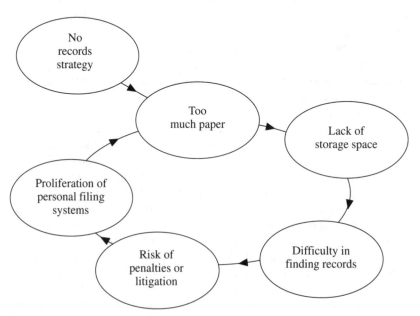

Fig. 3.4 The vicious circle of paper (*source: Touche Ross Management Consultants, Information Management Survey, 1992*)

It could be used to put systems or procedures in place which can be used in company marketing and perception/awareness campaigns. Maybe an EDM system could enable a much faster 'front end' cycle to identify and order spare parts for a particular company's products. They may well wish to use this aspect to gain a competitive edge over rivals by guaranteeing a fixed response to the delivery of spare parts where perhaps previously there was none. A promotional/sales campaign may be initiated to support this. Considerable market share could result from such an exercise but it must also be remembered that using technology in such a marketing promotion can be a double-edged sword. Remember the 'lead' Fiat cars generated a number of years ago with their 'built by robots' marketing message. This message, while being acceptable and leading-edge at the time, quickly lost ground to what we see now as a more 'hand-built/craftsmanship' image in the automotive manufacturing sector. It took Fiat a long time to shake off the image it had created for itself, so make sure the message generated by the use of technology is correct before using it. When EDM is successfully utilized to enable more efficient and businesslike internal procedures to exist, the end result will attract the market's attention, whether directly or indirectly.

The improved quality of product or service is a further aspect which should be considered when introducing EDM. This will of course prove to be of even greater benefit when considered as part of a TQM environment, but even on its own can provide significant benefits. This is mainly due to the ability of an engineering team to consider a number of alternatives within the given development timescales rather than accepting the first viable design solution, simply because of the inability of the engineering-related data to keep up with the momentum of the design process. Hopefully, the selection of the best alternative and freeing-up the design team to be productive in terms of the product and not in the headache of its associated paperwork will pay longer-term dividends for the entire process and the ultimate success of the products concerned. Enhanced product quality also arises from many of the other points already discussed thus far, including:

- Improved communications
- More effective use of existing proven designs and information
- Fewer errors
- Improved coordination and control

We can see how the previously discussed and now generally accepted aspects of:

- Reduction of lead times, by changing from a serial to a concurrent engineering environment
- The improved quality of product and/or service
- Getting designs right first time by re-using relevant data in a timely manner

can be gained as a result of a correctly implemented EDM solution. But so can one other vitally important function, which brings us full circle to our previous discussions. That is that they provide the ability to react to *change*. This will be one of the major differentiating factors in achieving success in the 1990s and beyond, and if there is only one point that is remembered as a result of reading this section, let it be that *EDM is a solution which can provide the necessary foundation to enable a manufacturing enterprise to more readily react to change.* Such changes, often identified by the growing acceptance and use of business process re-engineering (BPR) at the strategic level, require improved management planning and control facilities to be provided at all lower levels. These aspects can arise from, among others, the ability of an EDM system to:

- Make predefined and *ad hoc* enquiries of various kinds
- Report status information in a timely manner
- Link EDM to project management
- Identify problem products and take corrective action
- Provide full audit trail facilities
- Allow view-only access to relevant data

3.5   Mid-level and micro benefits

So far we have talked about benefits of one kind or another relating to EDM systems, but where are the cost savings? It is all very well presenting philosophical and strategic benefits to your board of directors, but the question, 'What is the bottom line?' will eventually be asked. This section will concentrate on the actual benefits (we have already considered some of the points about the financial justification of EDM systems and will consider this further in Chapter 6). Many people find the financial aspects to be the most difficult in justifying any form of IT spend, not just EDM, and so it will also be considered within each of the case studies in Chapter 10. Unfortunately, many of the cost savings which can be calculated are case specific, but given common baselines and a structured approach, can be suitably quantified. They are also now based around a more structured set of criteria. Happily the early days of computer system justification where the main thrust was in the number of people saved have been replaced by justifications based on increased efficiency and productive savings. This section will discuss the general types of benefits which can be gained.

Cost reductions

A reduction in overall costs during the design and development phases of a product can usually be established. This will change depending on the costing method adopted, but even in these cases where perhaps the reduction is small, a definite shift in the pattern should be able to be established from the previous predominance of indirect cost (e.g. paper chasing and inefficient use of time) to those of a more direct productive nature (e.g. minimizing design development costs by re-using data from previous projects of a similar nature). The use of an EDM system can also reduce direct warehousing and stock-holding costs. The correct administration of the database can minimize the number of outdated or duplicate/equivalent parts actually held. The ability to identify and implement a consistent use of all parts and assemblies can provide significant direct savings in this respect. These can be further enhanced by the correct use of a structured engineering change control process.

Fewer design errors

We are all aware that our chances of introducing design errors are increased by allowing the designers or the members of a product development team to operate under pressure in an unstructured manner, yet how many of us actively take steps to reduce this situation? Occasionally we can find environments where this very situation is actively encouraged to provide a stimulating and challenging environment: a 'My team work best under pressure' approach. While departmentally this may be true, the overall company-wide situation is usually best served by actively re-using existing data and providing a structured, organized framework to enable the creation of new ideas and products which are correctly thought out, from cradle to grave. An EDM solution can help to provide this framework. The types of problem which can result from an unstructured framework for new product development can be demonstrated by the following example. I was once employed by an engineering company which produced various types of valves, to the same original designs, for many years. They decided to design and produce a brand new range of ball valves and spent many months producing a fine range of models from 1/2" to 8" with a range of hydraulic, pneumatic and manually operated actuators. As usual the final stages of the new product introduction became rather tense with looming critical deadlines for product brochure generation and press launches, etc. One of the first models for final prototype build was the handle operated 4" ball valve, which when completely built could not be operated because the handle fouled on the valve flanges! The solution came in the shape of a highly cranked handle which looked ludicrous and completely negated the comprehensive design work originally undertaken. This, as it turned out, was the only viable alternative since the valve bodies could not be changed because the significant tooling investment required for their production had already been made. A number of lessons can be learned from this simple example (and they were!), not all of which require complex computer systems or EDM to solve, but which certainly could have benefited from a correctly installed CAD/CAM system, an EDM system or a CE/TQM environment. Fewer errors in design can result from correct procedures for checking and approving documents, correct access to released documents and less obsolete data for example—i.e. a correctly structured environment.

Better and more effective use of staff

We often hear that 'Our people are our most important asset', but how many organizations really believe that, or have found ways to effectively tap into that asset? Our aim should be to turn every employee into a problem-solver and an adder of value, not just someone doing a job. EDM helps by providing a foundation or framework, in that engineers can locate, retrieve, distribute and re-use data in a speedy and more efficient manner. This operates both within individual departments and on an interdepartmental basis, thus promoting teamwork and helping to remove both the 're-inventing the wheel' and the 'over the wall' engineering practices which may have become the norm. This, as we have discussed, allows engineers to be more productive and do what they were originally employed to do, not shuffle paperwork around an office.

Customer service

We have already seen how important service aspects will become over the coming decade. For example, the quality provided, the warranty and after-sales service, the provision of spare parts, etc. An EDM solution can assist here via the configuration management facilities provided which can be linked to or used in conjunction with suitable customer service functionality. This could be used to improve the cost/effectiveness of a spare part ordering system by suitable database interrogation to establish the build standard of the particular part and follow this through to the as-built BOM on the MRP system and so plan the despatch of the correct replacement part—all within one structured customer enquiry, where a number of such calls may have previously been required. A correctly administered CM system can also prove invaluable in a product recall situation by allowing the supplier to perhaps regain some credibility after a potentially dangerous product has been sold, by pro-actively recalling the product prior to any serious injury being caused. Once a product defect has been confirmed the units involved must be identified and located. To do this an effective system of product traceability must have been in force for some time. Product traceability has to be effective in two directions: upstream to suppliers and downstream to end users. How far along these lines traceability is maintained is a matter for managerial judgement. It is worth remembering that in some cases product traceability may be important to enable a company to show that it did not make a specific product. The costs of a recall situation also require to be kept in mind. For example, the recall of a tyre in the USA some years ago was estimated to be around $250m. This tends to focus the mind on avoiding them at source!

Change control process

Although traditionally an important aspect in any company's operations, change control is assuming an even greater role as we have seen. Most companies agree that even if their change control process is reasonably well defined and controlled, it could always be made better and this is an area where significant economies can be made from both a financial and operational perspective. Financially, an EDM-based change control process can help reduce the overall costs of change, by for example the more efficient utilization of staff involved in the process, less paperwork and costly reproduction techniques and the most efficient use of any outstanding inventory which may be able to be re-used rather than scrapped.

From an operational viewpoint, an EDM-based change control process can increase response from both individuals and departments and can increase effectiveness of the overall system by providing a secure proving ground for ideas to be developed before the work involved in the actual change is embarked upon. It can also provide a structured format for the adoption and use of a system which may be developed from the ground up or which may retain the use of existing methodologies or techniques if required.

Space saving
One final point which should be considered here is that EDM can save space. Many readers will be able to relate to the problems of working in a cluttered office environment or DO where drawings and documents are spread over every available flat surface or even stacked vertically in rolled-up form for possible future use. The savings in space which can be made by storing such data electronically can be immense and many DIP/Document Management systems have already been justified purely on this saving alone, especially in crowded city centre environments where space is a costly premium. However, although this is a tangible benefit which can usually be applied to most environments and easily cost justified, we must remember the strategic benefits which can accrue from using EDM and not reduce the potential of such a system to that of a mere electronic filing cabinet to tidy up our drawing office.

3.6   The intangibles
What is meant by an intangible benefit, and how did the term evolve? In the past many companies used the traditional payback method to financially justify an investment in AMT, MRPII, JIT, etc. and often found that they were unable to quantify many of the benefits they knew were available. When in this situation and unable to identify sufficient direct savings to justify the investment, they began to look for other benefits which were generally described in terms such as *improved quality, increased production flexibility* or *better management control*. The inability to show how such benefits could be quantified led to the assumption that they were intangible. We can therefore consider intangible benefits to be those which cannot be accurately measured—generally unquantifiable. Such benefits or factors could be

considered generic in nature rather than requiring specific proof figures to be calculated.

This view has persisted over the years and it has been generally accepted that there are truly intangible benefits which will result from investing in a particular IT solution such as EDM. These could include better resource utilization or increased security of data for example. However, there is a growing body of opinion which states that there is no such thing as an intangible benefit—the University of Manchester Institute of Science and Technology (UMIST) for example, which has backed up this view from practical work recently undertaken—and it is true to say that everything can be quantified given enough assumptions.

The usual nature of intangible benefits is that they normally occur in a different area of the company from that where the original investment was made, and their magnitude can only be estimated rather than calculated. Many EDM benefits are found further down the procedural/documentation chain rather than in the pure design/DO environment for example. This has led to confusion between the ability to put benefits into quantifiable terms with the accuracy with which their value can be estimated. It would appear that there are two distinct problems:

- The form in which the benefit can be quantified
- Estimating the magnitude of the benefit concerned

Intangible benefits can generally be broken down into a number of smaller, more quantifiable, elements. This is illustrated by an example from Dr Peter Primrose, of Total Technology at UMIST. He cites the generally accepted unquantifiable benefit of *better quality products*. This can be broken down into several quantifiable benefits, such as:

- Reduced scrap
- Reduced rework
- Reduced disruption of production caused by scrap or rework
- Reduced warranty and service costs
- Reduced need for safety stocks
- Increased sales of better quality products

By redefining the generalized statements in this way, the problem has now changed from the inability to include benefits, to the accuracy with which they can be estimated. This is a completely different and manageable problem where techniques such as sensitivity analysis can be used to improve the accuracy.

But once again the law of diminishing returns applies here in spending a disproportionate amount of time in specifying something over an extended timescale or with a large number of assumptions and a correspondingly increased margin of error, when the underlying principle is generally well understood anyway. Each case requires to be considered on its own merits with those benefits identified as being critical to an organization examined at the most detailed level, whether 'intangible' or not, while for others an indication of the result of the 'intangible' may be all that is required. Take, for example, the generally accepted benefit of an EDM system—that of improved communication. No one could argue that a correctly installed EDM system can improve communications between employees and departments. This comes about by many of the means we have already discussed and is also due to the increased ability of the system to distribute data by electronic means. For many years it was assumed that the use of computers would result in a paperless office, just as CAD initially promised a DO without a drawing board. We can now appreciate how misguided these views were and we should not repeat them by assuming that EDM will remove the paper element in our engineering environment. Nevertheless, there now exists the real prospect of at least increasing the proportion of data distributed by electronic means with a corresponding decrease in paper-based communication. These aspects of improved communication can in turn lead to improved teamworking, fewer errors, sharing of ideas, design process concurrency, etc. But attempting to detail the effects of the improved communication with regard to time and consequential cost savings could take a great deal of work and will probably ultimately prove to be inaccurate in any event. But improved communications at all levels in an organization is possible through the effective use of EDM.

A further benefit is that of increased operational efficiency. In the last section we discussed how EDM can assist in increasing the efficiency of both individual staff and associated departments. This concept can of course be taken further to look at the operation of the corporate level of the company itself, be it of a centralized or geographically spread basis. The concept of communicating design or engineering change information between a number of associated company plants in individual countries throughout the world may be a problem well known to many readers. Controlling the effective communication of such information within one physical enterprise is often difficult enough, so the ability to use EDM to increase the overall efficiency of a company's corporate operations enables a more reactive, or even in certain circumstances pro-active, environment to exist. This environment should enable the company concerned to react to change ahead of its competition and ultimately provide a significant competitive edge.

An EDM system can also ensure the security of information within a company via two main routes: access security to the system as a whole and data security within the system itself. This is especially required in a secure research and development environment or a Defence or Government restricted environment. These areas traditionally require full traceability and procedural capabilities to comply with the most stringent security requirements—all areas where EDM technology can be applied. Even within a normal engineering environment, the ability to define who can access information, what they can access and how they can access it can prove to be a significant advantage. Such aspects can be used to provide a secure proving ground to ensure correct product definition, refinement and testing before release.

EDM can also provide a structured environment to assist in the use of the correct company standards. All designers and engineers can have access to all current and historical part and structure data. As they all use the same data it is easy to apply standards for quality and for consistency in the use of parts and assemblies in both design and production.

We saw in Section 3.4 how EDM can be used to enhance a company's external image. There we were taking an example of consciously using the benefits of EDM, but intangible benefits can also be gained by the increased efficiency of the internal systems subconsciously projecting themselves on the external company image. This could be considered as being achieved almost as a 'by-product' of using EDM and the role of EDM as an element of a TQM environment can once again be seen—the marketplace gradually accepts the company as providing a quality service in every respect.

One final point which should be considered here is that of opportunity cost, i.e. the ability to accept tasks and business which would otherwise have to be turned away. Because an EDM system helps to increase the efficiency of the product design, management and change control process, we have seen how an increased number of alternatives can be evaluated in the time-span taken to previously evaluate one. By choosing the best option from a range of alternatives we should hopefully reap the eventual rewards (suitability of purpose, better quality, less re-design, etc.) rather than inheriting the problems that accepting the only solution offered may have ultimately presented us with.

The time 'created' by this increased internal efficiency could also be used to accept work which could previously not have been considered. This may be direct productive, sales-related work or internal tasks, which are usually the ones which suffer in a hard-pressed office environment, for example correct administration procedures, quality of departmental operation, filing and administration, research, training. While it is generally well appreciated that these types of activity need to be addressed, very often everyone is running flat out to cope with the given workload and these tasks either are neglected or addressed by additional staffing—areas which can perhaps be more easily costed to help justify an intangible benefit!

In order to bring our thoughts on the benefits of EDM to a conclusion, the summary Table 3.1 has been included. This is based on the justification benefits as originally detailed by Phil Newell of MSPL. The list has been split into two columns headed tangible and intangible in order to indicate those benefits which may require further analysis in order to fully define them and those which can be more readily justified. Also, while all tangible benefits will ultimately carry an associated cost, for ease of explanation some of the examples have been expressed as time.

Table 3.1   Tangible and intangible benefits of EDM

Tangible benefits	Intangible benefits
Time	*Management*
Reduced time to prove BOMs	Managing overall workloads
Improved productivity/less duplication of effort/less time spent searching for information	Managing engineering developments
	Managing and controlling engineering change
Reduced clerical effort	Managing execution of work packages
Electronic change approval, giving estimated 50% reduction in lead time for changes	Prompt actions through pro-active E-mail
	Reduced bottlenecks/better priorities
Electronic copy/compare of BOMs (estimate time savings)	Improved flexibility
	Better resource utilization
Turnover gains through reduced lead times	
Instant release of data to factory through electronic interfaces	*Information*
	Accessibility of design/engineering information
Design and production BOM views	Distribution of up-to-date information
Time/cost to prepare/respond to tenders	Ability to handle increased volume/variety of information
Cost	Part/drawing issue history
Reduction in manual filing	Product history
Reduced overhead of maintaining isolated departmental systems	Latest changes on all parts/documents
	Visibility of status changes/projects/ developments
Savings on company cost of non-conformance attributable to design and engineering, such as inventory BOM errors, EC BOM errors, customer return BOM errors, duplicate stock under different numbers, etc.	*Quality*
	Bill of Materials (BOM) integrity
	Company standards
	BS 5750/ISO 9000
Transaction errors/data inconsistencies	Audit trails
Reduced paperwork—the value of this will depend on the extent of the EDM coverage	Improved procurement policies
	Improved design quality
Removal of multiple entry of BOM data due to electronic interfaces with CAD/MRPII etc.	Conformance to CE standards
Part rationalization	*Security*
Warranty cost	Control of data access
	Security of information
	Legal
	Product liability
	Traceability
	Customer
	Increased service levels
	Part/drawing issue history
	Product structure issue history
	Access to BOM/project status if desired

4

Underlying database considerations

4.1 The database engine—the heart of EDM

Our deliberations thus far have used the term database many times and, apart from a brief description at the beginning of Section 2.5, the subject of databases has never been fully discussed. Due to its fundamental importance to EDM we will consider the subject of databases in more detail over the rest of this chapter before proceeding further to consider the other aspects of technology which influence EDM.

The information (i.e. the metadata) and associated references to and between this metadata which we can store within an EDM system have to be defined and controlled in some manner. A database system is the mechanism which is used to enable this information to be input, stored, updated, retrieved and reported on.

Let us try to position it first before we consider what it is. Some EDM systems are supplied with their own unique database, some are supplied in a form which links to a number of different generally available ('standard') databases and some are designed to operate in conjunction with one particular 'standard' database only.

The basic point to remember is that because the database system coordinates information and data from many sources, an EDM system cannot operate without one. A computer system configuration is often considered in a 'layered' fashion as shown in Fig. 4.1 to aid in understanding what can often be a conceptually complex setup.

Whatever database system is ultimately chosen it has to operate closely with the application software (i.e. the EDM system), the computer operating system and its associated file management and network structure.

The method of use and choice of database system used by the EDM solution can have a significant effect on the ultimate selection of that individual EDM solution by different companies. This situation is analogous to the one we have previously mentioned about the use of a centralized numerically con-

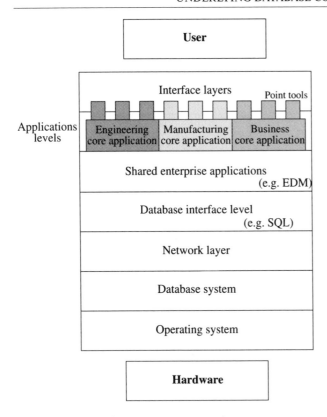

Fig. 4.1 A layered computer system environment

trolled (NC) tape preparation system. Each reasonably sophisticated NC machine tool purchased today will have a computer numerically controlled (CNC) controller provided with it to enable the direct input of geometry and program data to manufacture the part required—so why have a centralized NC tape preparation (CAM) solution? Well, the benefits are obvious and are being achieved by many companies, but in doing so they have had to choose between a number of options. For example:

1 Buy a machine tool with only a very basic controller to accept the instruction set generated by the centralized system (via paper tape or DNC).
2 Buy a machine tool with a comprehensive controller but do not use it.
3 Buy a machine tool with a comprehensive controller and use it in conjunction with the existing centralized system.
4 Scrap the centralized system and use all the individual controllers.

Choosing an EDM system which incorporates its own database can be considered to be similar to choosing the purchase of a machine tool with its own

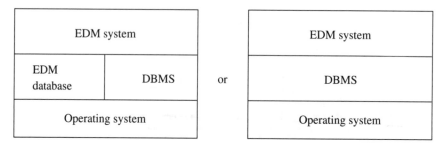

Fig. 4.2 EDM database options

controller. If you prefer EDM system X with database Y and your company has already decided its corporate database policy (and it is database Z) then you have a problem, especially if the EDM solution can only run on database Y. In such a case a decision would need to be taken regarding the corporate database policy or whether to adopt a systems integration approach and link the databases together in some manner. Such a choice, which is summarized in Fig. 4.2, is not easy due to its long-term effects on the entire company, not just the engineering-related sections.

We can see therefore how important the database element is—the true 'heart' of an EDM solution. Having considered its position within a system configuration, let us now look in more detail at exactly what a database is.

4.2   What is a database?

We all tend to use technology and its associated buzzwords very freely without really knowing what the names mean or what the associated technology does. In many cases we do not need to question it or understand it—we all use the telephone but we do not all need to know exactly how it works. The term database is one of these technology-driven words that we all tend to use often. But what is the difference between a datastore, a database, a database management system (DBMS) and an RDBMS for example? Do we need to know? Well, for normal everyday use, probably not. But as we have seen, a database is a fundamental part of an EDM solution's foundation and (perhaps unfortunately!) we need to understand it a little better. So let us start off with a definition. A database can be defined as *a collection of data which is shared throughout an enterprise and used for multiple purposes*, and from a corporate company perspective should be considered alongside other resources such as assets, stock, cash, and so on. This is represented in Fig. 4.3.

So what is the difference between a datastore and a database? Well, the key word in the definition of the database is 'shared'. A system which is designed purely for sharing data, a file management system, can provide comprehensive functionality but is usually considered in an isolated or 'compartmentalized'

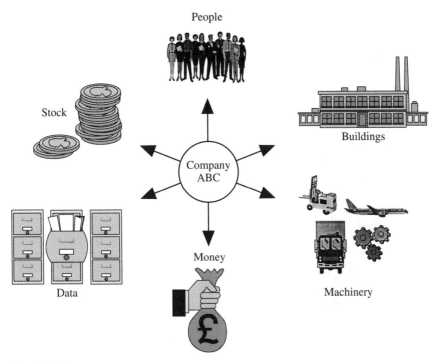

Fig. 4.3   Data as a corporate resource

way. This is shown in Fig. 4.4 where each application user creates a new file. We can see that such an approach is ideal for each user but a large installation of this type has hundreds or thousands of such files with a high level of data redundancy.

A database on the other hand acts as a separate layer between the user and the system itself to automatically store the data in the relevant place—the user does not need to know where, as long as the correct access and application program functionality is maintained. The important point to bear in mind here is that although a database is just a repository for stored information, it is now held in an *integrated* fashion to be *shared* and accessed by several users. In this case one file can be used by many applications with much reduced data redundancy (Fig. 4.5).

This concept, while providing many advantages in addition to pure data redundancy, has to be put in context and when considered in isolation is limited in its use. Some of the major benefits are obtained from the administration tools and functionality used to surround the primary database concept—the management tools—hence the term database *management* systems. Also, beneath the concept of the database and its associated management system is the actual method for storing the

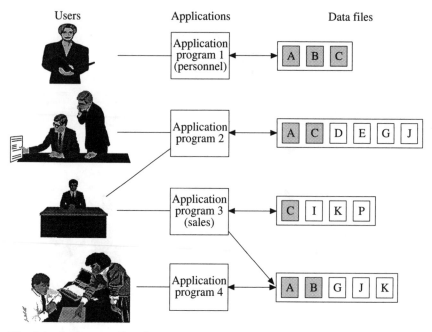

Fig. 4.4   A file-centred environment

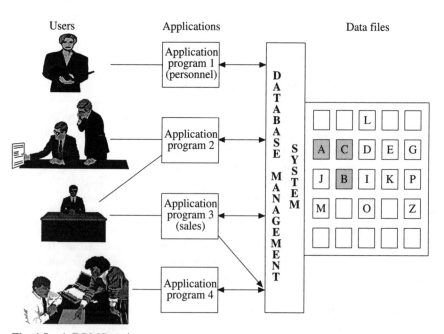

Fig. 4.5   A DBMS environment

information (or data). These are hierarchical, network and relational methods and are often referred to as DBMS 'models'. We will consider these further in Section 4.3.

We can see therefore that, just as we operate a manual filing system and retrieve only the data we require, the DBMS does not allow any one user to perceive all the types of data held in the database, but only that which is required at any one particular time for the job concerned. The information retrieved may be simple in nature, but could be derived from a much more complex structure—much like our own simplified use and view of our telephone. Thus a database is not only shared by multiple users, but is perceived differently by each of these users. In a way it is like taking part in that well-known party game where each participant is blindfolded and has to guess the nature of an object by touch alone. If presented with a zebra in such a scenario, one person may touch its leg and perceive it as being a tree. Another person may touch its skin and perceive it to be horse hair. Yet another touches its tail, and so on. This is similar to the situation where different database users perceive different *views* of the same data. The important point to remember here is that someone has to design the entire 'zebra' and in doing so must meet these often diverse needs of different users.

This is where the DBMS can help. Many different tools are normally available, such as access mechanisms to query and report on the database, structuring and optimization tools, application building tools such as fourth generation languages (4GLs), etc.—aspects which we will discuss in a little more depth later in this chapter.

From the simple view taken thus far we can see the uses to which a DBMS system can be put. Data storage is an essential resource for any business enterprise and databases can be found almost anywhere nowadays. Banks, insurance offices, taxation authorities, schools, hospitals, police, lawyers, garages, etc., all have databases and probably all now utilize a DBMS. The benefits of data accuracy, fast access, less data duplication, increased 'openness', etc., are well known and are currently being achieved by many.

So what is the difference between EDM and a DBMS? EDM, like many system solutions today, is written to take advantage of the DBMS upon which it is based. EDM is a database application to store and manipulate metadata which sits on top of a DBMS, as we have seen. So, we should consider a DBMS as an 'enabling' technology. It enables applications such as EDM, or perhaps a personnel system, order processing system or mailshot processing system for example, to utilize the underlying database and thus provide the quoted benefits. This acts in a similar manner to EDM itself being an 'enabling technology'—EDM provides even higher-level business technology to be utilized and the ultimate benefits realized.

4.3   Database management systems

The concept of the database has been in existence ever since man started keeping written records. Whether carved on stone slabs, written on paper and stored in a filing cabinet, or stored digitally in a computer, a database is simply a store of data. When related specifically to computer technology, however, the term database and its associated technology is often thought of as a relatively new field, but the origins of the database as a 'computer' data storage technique can be traced back as far as the late 1800s with the invention of punched cards. These remained the leading method of storing data for over sixty years, and continued after the introduction of the first electronic computer, since these early machines were mainly used for scientific applications.

A database management system (DBMS) extends the concept of a simple database to include the administration tools and associated functionality which surrounds it. These have grown from the early use of computer-based databases and are an important aspect which are still undergoing rapid development.

Initially, magnetic tape, which became available during the 1950s, had a dramatic effect on data storage even though it was still a sequential storage medium. The terminology associated with punched cards such as files, records and fields continued to be used with the newer magnetic tape systems. Similar effects were felt with the introduction of the magnetic disk in the 1960s, and it provided a direct access storage mechanism to allow database management systems to support the integration of stored files. At this time, it was realized that what was needed was a general database system to store this integrated data and serve a range of applications (i.e. be both program and language independent). It was also realized that changes to this data should not require the associated applications programs to be changed and that the overall system would need to contain data representation facilities together with a range of data access techniques.

The original aspects of database technology have, like other areas of IT, moved ahead in acceptance and ease of use, with the original 'black box' status replaced by a generally available computer-based tool which can be utilized from a basic PC to a large mainframe installation.

Before we move on to look at specific database types (or 'models'), let us consider some of the terminology, technology and objectives of DBMS. In general, DBMS technology has three main objectives which are all based around the concept of hardware independence. The main aim is to bring information systems development closer to the way people work, rather than forcing people to work in a way dictated by the computer.

- *Physical independence*   This is provided at a lower level by the computer's operating system itself which, together with the DBMS, provides the programmer with independence from tedious details such as record and data field breakdowns, disk blocks and storage details, etc.

- *Access independence* Each application program needs to contain the necessary access paths to the data concerned and the objective of a DBMS is to remove any unnecessary details here. For example, the database query language does not require *how* to access a specific record to be specified, only *what* data is required needs to be detailed.
- *Data independence* One of the main aims of a DBMS is to conceal database changes from users or programs except those that need access to the new data items, thus removing the previous dependence on particular methods of access to data.

There are many other aspects of DBMS which should be briefly mentioned. These include the ability to reduce redundant data, although this is not an automatic benefit since it normally relies on the programmer correctly utilizing the appropriate tools. A database administrator will generally be required for any sizeable DBMS to organize the data in such a way that all applications share it correctly. This is an important position since it requires an understanding of an entire company, not just individual departments. Database administration cuts across all application boundaries.

A DBMS also uses certain data and file structuring techniques. Briefly, these are:

- *Data records* A data record is a meaningful object about which there is a need to record information, for example, accounts, customers, drawings, suppliers.
- *Data attributes* A data attribute is a component of data stored against the record itself. For example, attributes of an engineering part 'record' could be part number, quantity, description, etc.
- *Files* Files are named collections of records stored and organized in such a manner as to maximize their eventual retrieval. Types include serial, sequential, random, indexed.
- *Relationship* A relationship is an association between records which may be required to provide the logical operation of the database system. For example, assemblies *comprise* parts, parts *appear on* drawings, etc.
- *Data dictionary or 'schema'* This provides the description of the data architecture within the database itself, for example the individual data types, descriptions, lengths and relationship and storage details.

These concepts can now be utilized in the three main database models which are outlined below, together with a brief mention of some other models which may be encountered.

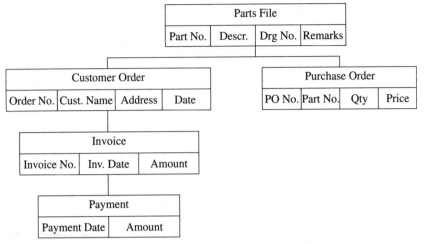

Fig. 4.6 A hierarchical representation of a simple set of engineering data

The hierarchical model

This model was the first to be implemented and, as the name implies, requires that the data be arranged in logical hierarchy with each data element containing a pointer to only one element in the level above it. This is often represented as a tree structure in a similar manner to a BOM (see Fig. 4.6) and a complete database can consist of many separate structures. Many applications have been developed which utilize the hierarchical model and the technique has proved very successful. However, there are a number of aspects which have reduced its popularity over the past number of years. The main problem is that it is rare to find a complete set of data that is purely hierarchical, even though certain parts may be. Imposing a hierarchical structure over all data can lead to serious problems and unnecessary coding complications. The model also has difficulty in expressing a 'many-to-many' relationship, rather than the 'one-to-many' relationship as outlined and can be inflexible and difficult to alter.

The network model

The network model does not require to be tree-structured (Fig. 4.7). The network model overcomes the problems encountered with the previous model by viewing the data records as sets together with their associated members. In order to describe the set associations, the network database utilizes a 'connector' record, which links records together to form data sets. A network database model does not allow a record type to be both the owner and a member of a set, so modelling a hierarchy where perhaps an assembly can own another assembly requires a level of 'indirection' to be introduced, which in the normal

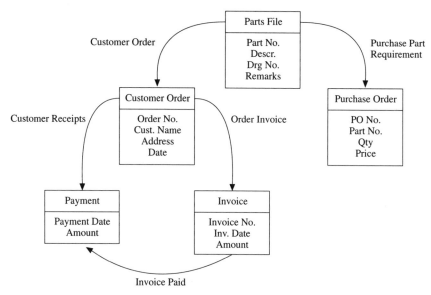

Fig. 4.7   A network representation of a simple set of engineering data

course of events should not prove to be excessively difficult. Unlike the hierarchical model, this model does allow the user to store and retrieve many-to-many relationships in an efficient manner and is more flexible, but in general requires a more complex data dictionary or schema together with detailed data access procedures. This model has been adopted mainly for large database systems such as finance and business institutions due to the modelling specification on which it is based, known as the CODASYL model, being closely linked to the COBOL language and conceived by a task group from the COBOL committee. It also has a higher level of performance than the hierarchical model. Some engineering applications which utilize the network model can be found, but they are more of an exception than the norm.

The relational model
The relational model has proved very successful over the past decade. One of the main reasons is its inherent simplicity, as can be seen from Fig. 4.8. The theoretical model was developed by Dr E. F. Codd, and, in the early 1970s, gained popularity through its initial uptake by C. J. Date in his book *An Introduction to Database Systems*.

The relational approach to the storage of data is to view data as 'relations'. This is a mathematical term for a two-dimensional table, which in the case of a relational database would be a collection of records. Commercial products use obvious terms for the database: record types are *tables*, records are *rows* in

Parts Table

Part No.	Descr.	Drg No.	Remarks

Customer Order

Order No.	Cust. Name	Address	Date

Purchase Order

PO No.	Part No.	Qty	Price

Invoice

Invoice No.	Inv. Date	Amount

Payment

Payment Date	Amount

Fig. 4.8   A relational representation of a simple set of engineering data

a table, and fields are *columns*. The academic terminology, however, is somewhat different: tables in relational terminology are called *relations*, rows are called *tuples*, and columns within a record are *attributes*.

Note that there are no predefined connections in a relational database as with the previous models we have discussed: the fields which require to be connected are simply duplicated and the actual connections are done dynamically. This dynamic joining capability means that the structure of any individual table can be modified without affecting the remainder of the DBMS or existing application code. These aspects of the relational model mean that it is ideal for implementing application solutions in a modular fashion, thus enabling new solutions to be introduced quickly and their functionality improved in an incremental manner.

One of the major contributions of this model is the provision of a powerful 'query language' which provides the user with a direct method for easy data access. This can be embedded within the application code itself or used on its own as a 'high-level' interaction language. The two most commonly used are the Structured Query Language (SQL) and Query By Example (QBE). SQL was originally developed by IBM and has now gained international recognition as the leading relational query language with the syntax itself now forming an ANSI (American National Standards Institute) standard. The operation of SQL is represented by such statements as SELECT, FROM and WHERE. For example, if we wanted to list the parts supplied by XYZ Company, we would type:

SELECT "Part Numbers"
FROM "Suppliers"
WHERE "Supplier Name"="XYZ Company"

QBE is a two-dimensional interaction language which requires the user to complete columns on a representation of a table thus forming the example query. This is usually done by function keys to enable the technique to be used on a wide variety of system screen devices.

Relational database management systems (RDBMS) are now widely used in almost all application areas due mainly to their simplicity and user flexibility. The ability to dynamically add new relationships to existing data overcomes one of the greatest drawbacks in the upgrading of application software. The relational model has been known to prove slower in use than the previous two models but recent hardware and software technology developments have reduced this limitation considerably.

There are many other technical developments which have had a significant effect on the use of DBMS; two of these are mentioned below to act as a suitable introduction to the subjects. Further information on database technology can be found in the references included in the Bibliography at the end of this book.

Object oriented databases (OODBMS)
This is becoming a more widely accepted concept in database technology and is simply the term used to describe a collection of integrated applications and their associated data. The 'objects' themselves are merely data items of a more tangible nature such as a 2D drawing, a scanned image, a voice message, etc., which in turn rely on the dependent applications, which are controlled by a central database application or 'executive'. The object oriented database provides the executive with information and it decides where the information is stored and which individual application should be run to process the data (Fig. 4.9).

Distributed databases
Distributed databases are simply defined as collections of data which belong logically to the same system but are geographically spread over a computer network. The benefits and associated technologies of Local Area Networks (LANs), Wide Area Networks (WANs) and client/server architectures (CSA) are generally accepted in today's computing environments, and like other aspects of IT have conspired to affect other related aspects of computing, with database technology being one of the prime areas. In addition to these technological factors, the distribution of data has also become an increasingly important area in information processing over recent years for a number of

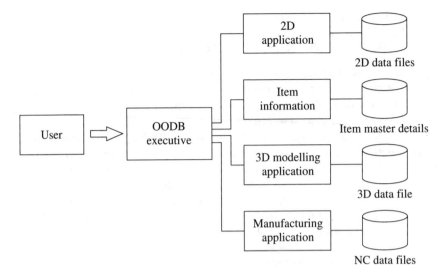

Fig. 4.9 A representation of a simple OODBMS

organizational reasons. For example, we have seen a move towards more decentralized organizational structures which naturally fit distributed data-base systems and the approach can similarly provide a smoother growth path when adding new units to an already established organization.

4.4 The need for data modelling

It is possible to purchase computer programs and applications directly from a high-street computer store. These are generally known as 'shrink-wrapped' software packages. Popular examples include computer games, word-processing, spreadsheet and business graphics packages and are generally of a PC-based nature to run on MS-DOS machines. It is generally unusual for larger mini or mainframe software to be sold in this way. Why?

There are many reasons, most of which are outside the scope of this book. What should be mentioned here is that these larger software packages are purchased for strategic, often 'mission critical' applications such as an accounting system or manufacturing resource planning. As such they require a much higher level of support and supplier knowledge than can be provided by a high-street retailer and almost always require a great deal more effort to set up and implement. In general, they require someone with a knowledge of the application area to 'mould' them to the environment under consideration. In some cases this may involve 'tweaking' (slightly modifying) the system via standard facilities (or extra custom code) to perform exactly what is required and as a result they need an increased level of training and support to ensure that they meet the objectives initially set.

A DBMS is a prime example of such a system. Smaller PC-based DBMS can be simply purchased and, after suitable operator training, can be put to productive use. However, the purchase of a fully functional, comprehensive DBMS such as one of the leading RDBMS (Oracle, Ingres, Informix, etc.) is a much more serious undertaking and is often a strategic decision, as we have already discussed. Usually, these systems can be purchased in modular form (in terms of both functionality and the number of simultaneous users) and in essence consist of a 'toolkit' of software routines which allow individual applications to be created. In some cases a 'core' or partial application is provided with the 'toolkit' to enable a faster take-up of the eventual application, and indeed some EDM systems are supplied in this form.

Thus, it can be seen that a much greater amount of 'up front' work is required with such a database system—you cannot purchase it on day one and have it working on day two. The writing of the application is a key area in the successful use of any DBMS. This will involve one or more business or systems analysts preparing a series of specifications to define the requirements of the application from an operational perspective, together with the necessary details on how the DBMS will be tailored to address these from a technical viewpoint. In many cases the services of an external consultancy or systems integration specialist is sought to address these issues. Once this phase (or series of phases) is complete and the application is ready, the users can be trained and 'live' use made of the system. This process is summarized in Fig. 4.10.

In a way this situation is analogous to purchasing a new suit. We may be faced with the choice of buying a suit from our high-street store 'off-the-peg' and accepting the standard sizes provided or we could have it made especially for us by visiting a suitable tailor. The tailor will note our requirements in terms of cloth, colour and design, style, sizes and linings, etc. (i.e. a specification) and will subsequently plan how to make the suit (a functional plan). He will then undertake the task (i.e. develop the suit) and request one or more trial fittings to refine his work (i.e. test it with regard to quality, size, etc.). Once complete it is delivered (it is accepted) and worn as required.

The use of DBMS technology is like having a suit tailored for you. You get exactly what is required and, although it may cost more initially, the overall

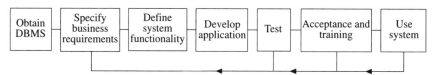

Fig. 4.10 A typical DBMS application development cycle

business benefits should far outweigh this (if you get your sums right!). The alternative is to select a standard software package (an 'off-the-peg' suit) and accept its functional fit. This stage, whether via standard or custom software, is an ideal time to examine some of the procedures which the system may be required to emulate. Perhaps these have been established over many years with no clear reasons or logic as to why they should continue. Now would be an ideal time to incorporate procedural and/or cultural change before coding the database and accepting an application designed to suit an inefficient set of working practices. This is discussed in more detail later in Section 7.1.

This aspect of defining requirements is often termed 'system modelling' or 'data modelling' and is used to verify the overall application in a structured manner prior to live use. There are different levels in this modelling process (hence the different terms) but it must always be remembered that at the lowest level it is data that is being dealt with and that data items do not exist in isolation but are associated with each other in some fashion. We therefore need to map or draw which of these data items are associated with each other, together with the type of association itself. Data models provide this representation of the data that is required to be controlled by the application concerned. Modelling data is a fundamental aspect of successful database operation—there are many database installations where it has not been done adequately and the full advantages of database use are consequently not achieved. Data definition and modelling go hand in hand and there are many different methods of defining the logical data structures and definitions required. The reader is referred to the Bibliography for further reading if required. The main point to remember in any modelling system is that it should provide a clear representation of the data via diagrams and sketches. These can then be given to analysts and users to check, refine and provide the necessary feedback. A simple example is shown in Fig. 4.11.

One of the main aspects of incorrect data modelling is in the longer-term maintenance of the application code initially generated. In many database applications more than half of the money spent on development is being used to re-write that which is already written. The euphemism 'maintenance' is often used to describe this situation. It may be that a new application suits the

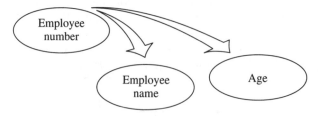

Fig. 4.11 A simple data model representation

Expenditure

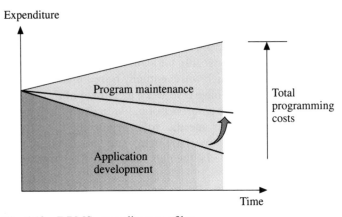

Fig. 4.12   DBMS expenditure profiles

user's needs initially but as time goes by and unforeseen changes are required the original structure, if not designed correctly, may become unable to cope. This generally results in a greater amount of resource being spent in maintenance (i.e. restructuring and changing existing code) rather than in generating new functionality. While an element of both will always be required, it is preferable to minimize non-profitable maintenance and maximize any new functionality which could be profitably utilized. This is shown diagrammatically in Fig. 4.12.

4.5   Administration and control

There are a number of popular misconceptions about DBMS:

1   A DBMS will not normally contain *all* the data belonging to an organization. Certain people still believe this to be so, but in general a company or organization will initiate DBMS technology in one area initially and 'seed' this into others.

2   Similarly, there is a belief that an organization will have *one* database. Again this is unlikely to be true, as we have seen.

3   A database or DBMS does not necessarily imply an MIS (Management Information System). The use of the word management refers to the management of the data, not the use of the database by management.

4   Once the application and data have been analysed and modelled and the application is running live the DBMS does not maintain itself. It requires adequate maintenance and administration to ensure its continued successful use and future development.

Although all of the above points are important, it is the last we will discuss in greater detail in this section.

Computerized data within an organization is not static. Like any manual filing system, the details of the data stored and its use are changing continuously in a dynamic and fluid manner. Therefore the structure of any DBMS requires to be similarly modelled—an unchangeable file structure will not be able to respond to the necessary changes which will eventually be required of it.

Many larger organizations which have invested a great deal of resource in database system technology will have a number of individuals within their IT or DP departments who will have specific roles regarding the DBMS employed. Titles, responsibilities and roles are many and varied but, in general, the following needs to be provided if the maximum return on investment on the database is to be achieved:

- *A data strategist*   To manage corporate data requirements.
- *A database administrator*   The person who presides over the process of defining and modelling the data.
- *A database designer*   The person who does the physical database design.
- *System analysts*   Persons who provide the necessary input to the data administrator.
- *Database programmers*   The persons who code the application software according to the agreed specifications.
- *Database operators*   Persons who ensure the availability of the system as and when required (i.e. maximize 'up-time').

The overall structure is diagrammatically represented in Fig. 4.13.

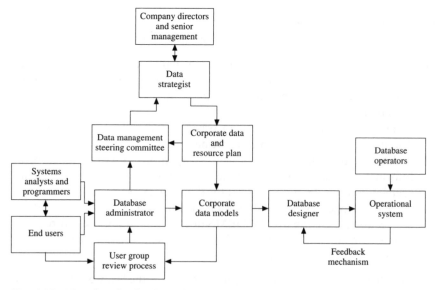

Fig. 4.13   Planning the database development process

Any database system requires these roles to be filled if it is to be successful. However, many smaller organizations will be unable to justify a separate person for each of these roles and will normally combine them as required, often down to one person in the smallest of operations!

Certain tasks require to be performed, whether by one or a number of people. The main ones are noted below and are not just EDM-specific—they are generic database administration tasks which are required on any DBMS application.

Maintenance and enhancements

These occur on two fronts in a DBMS environment. Firstly in relation to the DBMS itself where perhaps a new release of software is provided or additional requirements are established. A DBMS, although considered as a 'toolkit', is subject to the same requirements as any other computer software—namely new releases, upgrades, operational work-arounds or enhancements, etc. These require to be planned and implemented where necessary. The second area is in relation to the actual application itself—this will also require code maintenance, testing and upgrades which can also interact with the DBMS releases. Both of these areas require a considerable administration overhead: they require a quality approach, with planning and configuration management techniques utilized where necessary.

Fault reporting

Once again this aspect relates to both the DBMS and the application itself. A suitable procedure requires to be put in place whereby an application user can note down any bugs, problems or wishes regarding the application which can be submitted for action. Similarly any faults or problems regarding the DBMS itself have to be handled in an efficient manner (i.e. tested, worked-around, noted to the DBMS supplier, etc.) and may then result in the loading of a DBMS software upgrade or new release as noted above.

Related technological improvements

We are all aware how quickly technology can change. One aspect of the database 'guru's' role is to keep abreast of these changes and select those areas which could be successfully utilized. This could be hardware related such as an improved database 'server', or the use of affordable optical disk technology, or could be software or network related. For example, it could be the addition of a complementary software system which could be used to ease the generation of statistical information directly from the database itself.

General administration

It is a generally accepted fact that within a few months of using a DBMS, the value of data to the company which is stored in the database far outweighs the capital cost of the DBMS itself. It is therefore vital that it is maintained correctly. This means establishing proper procedures and tools to ensure:

- The *integrity* of the data—ensuring that it is not corrupt and can be used with confidence.
- The data is backed-up properly. Should the entire system fail or 'crash' a recent backup is usually required to reload the system up to the stage immediately prior to the point of system failure. Most systems will provide either 'roll forward' or 'roll back' procedures to achieve this. Backing-up a system at regular intervals is therefore essential.
- Archiving data. This covers the aspect of storing data (mainly on tape) if its live use is infrequent, and recovering it in a speedy and efficient manner when it is required—again an important task and one where each DBMS will generally provide automated software and administration tools to ease this job considerably.
- Security of data. A DBMS often contains highly confidential data. It is therefore important that such data is kept secure and not divulged to unauthorized users, otherwise confidence in the system will diminish to such an extent that it may never be fully recovered.

Networking, distributed databases and CSAs

These can be used to complement the structure of an organization and how it changes and adapts itself to suit a more fragmented, distributed or perhaps centralized nature depending upon both individual market and more global business considerations.

In summary then, DBMS and related technologies must be constantly reviewed. It is absolutely essential to plan for uncertainty in a DBMS environment. It must be possible to allow different systems to develop in their own ways, and as they evolve, they should employ data, technology and services which have been defined and modelled appropriately.

5

Open systems and related computer technologies

5.1 The background to today's computing technology

Like many major computer-related and IT projects, the saying, 'Computers make it possible but people make it happen' can be applied to EDM. Historically speaking this saying has of course always held true but has become more prominent over the past few years, especially when applied to specific techniques such as Manufacturing Resource Planning (MRPII). When the original technology of MRP (Material Requirements Planning) was introduced by Joe Orlicky and began to become more widely used and accepted, it was originally seen, quite correctly, as a computer application to address the fundamental manufacturing equation as originally advocated by Oliver Wight, which is:

- What are we going to make?
- What does it take to make it?
- What have we got?
- What do we have to get?

and was originally used in a simple manner, primarily as a structured method of ordering parts. Because of this the focus could be placed on the fact that MRP was a computer-based priority planning technique.

However, the early 1980s saw Oliver Wight introduce the concept of MRPII which, as well as including MRP itself, consisted of capacity planning, MPS (master production schedule), purchasing and shop-floor control—all operating on the same common database. It was also realized that accounting data could also be accommodated and it was therefore included in the MRPII concept. MRPII, therefore, developed from the practical use of existing tools and extended the concept of closed-loop MRP into a planning tool for all the resources of a manufacturing company: manufacturing, marketing, finance and engineering.

In many cases the original view of MRP as a computer-centred system was retained and applied to MRPII giving an incorrect perspective and contributing to the risk of such systems failing to achieve their maximum potential. Because of the changes to the original concepts and scope of functionality, MRPII has now grown into a full-blown simulation tool for a manufacturing enterprise and consequently the emphasis has changed from being a computer 'system' to being a 'people' orientated solution which happens to use computer technology to speed up the ability to process the required information.

The correct adoption of a modern IT solution such as an MRPII system requires a correctly structured and balanced approach, top management involvement and commitment, the necessary levels of awareness, training and cultural change within an organization as well as the necessary technical focus on the hardware and software to suit the process concerned. This has been proven in practice and very often learned the hard way by some early (and some more recent) implementations. This aspect of the technology acting as an enabler to people and processes is shown in Fig. 5.1.

Can we draw any parallels between this situation and EDM? Yes, we can—there are many similarities. However, we are not suggesting that implementing an EDM solution is on the same level as implementing an MRPII system. The introduction of an MRPII system into an organization is a complex undertaking which if handled incorrectly can bring the company concerned to its knees. An EDM system requires a similar process to be adopted with similar long-term planning, awareness, training and people-orientation, but should rarely have the same potentially devastating effects on the enterprise as a whole if handled incorrectly.

While both MRPII, EDM and the majority of large IT projects now recognize the importance of the people who will implement, manage and use it (and these are aspects we will consider in greater depth later), the computing aspects associated with the technology concerned cannot be ignored. We will therefore consider some of these in this section in overview form. Such information has not been included with the intention of providing a fully detailed

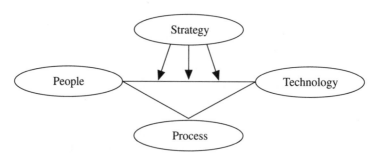

Fig. 5.1 The success characteristics of business integration

historical account of the development of the digital computer. It has been included merely to provide a background to certain of the milestone events which can be identified as helping to form today's computing environments, which in turn enable the successful adoption of solutions such as EDM. These include:

- *Mainframes and storage mediums*   Large centralized computing environments. Air conditioned computer suites. Traditional core memory—drum and disk storage devices. Strategic application environments.
- *Workstations and mouse-driven interfaces*   More widespread use of CAD and CAM. Ownership of resource and data. More powerful machines. Office environments.
- *Graphical applications and software developments*   Computer graphics art. Simulation graphics. 3D modelling and colour shading. CFD/FEA, etc.
- *Database and operating systems*   Codasyl DBMS. RDBMS acceptance. SQL. Object-orientated DBMS. Client/server architectures (CSA). User tools. UNIX and C.
- *Increased power and reduced size*   Ten-fold increases in CPU power. Reduced purchase cost. Cost of ownership. 'Footprint'. Mainframe/ mini/workstation/PC/laptop/notebook distinction blur.
- *PCs*   Basic introduction as 8-bit home computer. Developments and refinements. Dramatic effect on computing history.
- *UNIX*   The move to 'open systems'. SUN Microsystems. Apollo computers. C. Standards bodies. Consolidation. Acceptance of open computing.
- *Window environments*   User interface (UIF) technology advancement. Standards such as Phigs, GKS, etc. X-Windows. Motif and OpenLook window managers. Microsoft Windows and NT.
- *Coding developments*   Cobol. PL/1. Fortran enhancements. C. Object languages. C++.
- *Viruses/control/management*   Hacking. PC control and networking. Viruses and software copyrights. Dongles and software licence technologies.
- *Networking and distributed computing*   CSA. TCP/IP. Ethernet and token ring. Move from centralized computing. Downsizing and rightsizing. The rise of groupware applications and workgroup computing.

5.2   Open computing—a management overview

The emergence of open systems has given way to a new era in information technology and its effect on business practices and operations has been, and promises to continue to be, dramatic. The removal of many fiscal and legislative barriers means that many new business and geographical markets have opened up. The ability to communicate freely and exchange information in all areas of business has therefore become a commercial necessity.

This is where open systems come in. Open systems refers to the harmonization of information technology systems. By introducing homogeneous standards, open systems aim to solve the traditional problems of system incompatibility and allow business operations freedom to work across trading and operational boundaries.

In the past, manufacturers built systems which ran only their own operating systems and their own applications software. Companies which installed such systems found themselves locked into a proprietary spiral and unable to communicate outside their own closed environment. Only application programs written specifically for their system could be used: if they changed their computers, the programs had to be rewritten; an often expensive and inefficient procedure. The following quote may be typical for many organizations: 'Like many organizations, we developed our computer systems over a period of time to address particular operational requirements of our business. Now we have a variety of equipment and systems which need to exchange information between departments, sites and other trading partners. We have faced great difficulties and high costs in creating the required connections between these different systems.'

Open systems allow these types of situations to be successfully tackled. They open up the computer market to let all types of equipment work together, communicate and share information. The arguments in favour of such an approach are overwhelming, both in terms of logistics and economics. Most major computer suppliers are committed to adopt internationally agreed, homogeneous standards, mainly as a result of user demand—any new IT initiative must conform to these open systems standards if it wants to survive. This ensures that, as a user of such systems, you have the freedom to source the best equipment and software on the market from any chosen supplier, and at the best possible price.

So, for the purposes of this book, the following definition of open systems will be used, as taken from the book *Practical Open Systems—A Guide for Managers* by Ian Hugo:

> Open systems are those that conform to internationally agreed standards defining computing environments that allow users to develop, run and interconnect applications and the hardware they run on, from whatever source, without significant conversion cost. (*Reproduced with the kind permission of Data General Ltd*)

The concept of open systems is blindingly obvious. It is all about standards: standards in the way computer systems communicate and work together, standards in the way application programs interface to equipment and standards in security and quality. We take standards for granted in everyday life. For example, we confidently expect that any make of light bulb will fit into a lamp holder or that any electric plug will fit into a socket. If the electric sockets in our house only accepted one specific make of expensive plug, we would be annoyed, and rightly so. Standards have increased our freedom of choice and save us time and money.

However, information technology has developed at such a speed that relevant standards are only now catching up. This has been driven by the practical use of existing computer technology and in many cases has caught a number of suppliers (and users) out, due to its rapid uptake. Now all major suppliers are committed to providing some open systems products and plan to implement many more in the near future.

The most complete open systems standard is OSI (Open System Interconnection) which was developed by the International Standards Organization (ISO) and lays down guidelines regarding how different computers communicate with each other, via appropriate communications network design. The introduction of sophisticated data communications into computing has opened up a whole new era in application technology and allows companies to communicate locally, nationally and internationally, regardless of what machines they own. Several user groups and organizations have set up functional profiles or procurement specifications based on OSI standards. These include MAP (Manufacturing Automation Protocol) initiated by General Motors, TOP (Technical and Office Protocol) initiated by Boeing for technical office use and GOSIP (Government Open Systems Interconnection Profile), the procurement standard for office administration pioneered by the UK CCTA (Central Computer and Telecommunications Agency). Electronic Data Interchange (EDI) implements internationally agreed message standards. Applications portability is another area where standards are vitally important. This is a concept that allows programs, data and people to work with any make of computer hardware. In general, software applications have a far longer life-span than the equipment on which they run. This represents a considerable investment in resources and training, both for the developer and the user, whereas computer hardware is dropping in price and improving in functionality on almost a daily basis. The need for portable software has led to the POSIX (Portable Operating Interface for Computer Environments) set of standards which are specified by both UK and US Governments as part of their procurement definitions.

Appropriate standards for security, software quality and quality management are currently being developed to provide protection for the confidentiality, integrity and availability of data and better software.

Open systems standards are creating an intensely competitive marketplace. Computer system vendors can no longer depend on a captive market and must now compete on performance, price, reliability and service.

So what are the practical benefits of open systems? For many organizations the key to future business success depends on their ability to develop and use effective information systems—business operations and IT have to develop together if we are to ensure a profitable future. A number of major benefits can result from open systems:

- *Total freedom of choice*   Software can be chosen which meets the exact needs of the user from a wider number of suppliers and manufacturers. This can be done in the confidence that it will be readily available and will work together with existing equipment (i.e. be compatible) in one smoothly integrated unit, across all trading and operating boundaries.
- *Freedom to grow*   No business is static and computer systems must be able to change and grow as circumstances alter. Systems therefore require to be scalable—to allow freedom of movement between different sizes from desktop to mainframe. Open systems standards provide a stable base for expansion and future technical development.
- *They provide value for money*   Because open systems allow potential users to shop around, initial capital expenditure on software and hardware can be significantly reduced and greater systems compatibility reduces the cost and time for installation and training. Data communications are easier, cheaper and simpler to maintain. The aspects of portability and interoperability also contribute to better price/performance.
- *Greater efficiency*   Fierce competition between manufacturers, spurred on by user demand, has created open systems technology that is fast, powerful, functional and value for money. By utilizing these elements of open system solutions, significant improvements in efficiency can be made in many business environments.

Many people wonder what constitutes an open system and how they would recognize one. This is often not as easy as it sounds since, in many cases, the interpretation of what is open can vary, although as applications which rely on open systems standards become more accepted, the understanding of what is open and what is not will correspondingly increase. A further source of confusion can also be attributed to the word 'open'—just because a system is open does not mean that it is not secure. On the contrary, the importance of making appropriate provision for security when developing standards is well understood although it must always be borne in mind that it is a management responsibility to ensure that adequate security measures are adopted since the standards and products alone cannot deliver security.

Unfortunately, a trip through the relevant standards and standards organizations related to open systems is likely to induce glazed expressions since one of the most bewildering aspects of standards and standards organizations is their sheer number and diversity. This has often caused considerable confusion to those normally outside the IT arena and has, in some cases, done more harm than good. However, the proliferation of standards in the computer industry is no more a problem than that of, say, the building industry where there are standards relating to materials, plumbing, safety, insulation, electrical aspects and so on. The main difference is the fact that in the construction and building industry most of the relevant standards are already defined. A number of the main standards organizations are described in Section 5.7.

The adoption of an open systems approach to computing requirements can provide significant benefits as we have seen, and the trend towards its adoption has increased dramatically over the past few years and shows no signs of slowing down, as can be seen from the 1993 Price Waterhouse/Computing Opinion Surveys on open systems policies (Fig. 5.2).

However, many of these IT-related technologies and philosophies, just as with engineering-related topics such as MRPII, EDM or SFDC, have to be considered in a real-world environment. The theoretical hype needs to be cut through and each case for open systems considered on its own merits. Open systems are supposed to emphasize business issues and free users from the constraints of particular vendors and technologies. There are many success stories relating to the benefits actually achieved from the adoption of open systems, but equally so there are systems which have not lived up to their full potential for a variety of reasons. One thing is clear, however: any organization embarking on a long-term IT strategy must consider the adoption of open systems, even if they are ultimately discounted as being unsuitable. The potential benefits can prove substantial over the longer term and many board members will require to be convinced that the subject has been given the consideration required before agreeing a substantial IT spend.

5.3 EDM and open systems

Having now discussed the nature of EDM, database technology and open systems, we can start to appreciate the common underlying themes of information sharing, distributed data access, client/server computing and the need to plan an 'open' strategy to ensure the provision of adequate computing technology over the coming years. Although the concepts of EDM have been understood for many years, and indeed often applied using manual procedures and paper-based solutions, information technology has now progressed to the stage where true enterprise-wide computing technology such as EDM can be successfully adopted.

While early EDM systems were proven to be successful productivity aids, the related technological aspects held them back from achieving their full potential. This situation need no longer be the norm and fully configured computer systems such as that shown in Fig. 5.3 are now actively supporting those technologies such as EDM.

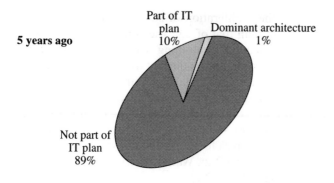

5 years ago

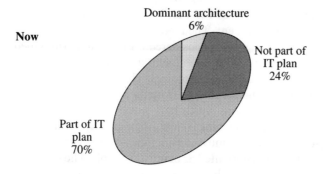

Now

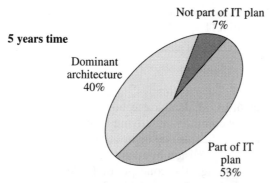

5 years time

Fig. 5.2 Open systems policies (*source: Price Waterhouse Computing Opinion Surveys, 1993*)

The adoption of open systems has grown from a number of different requirements and is closely linked to the development of the UNIX operating system—many people still say open systems when they mean UNIX and vice versa. Whereas the need for an open systems approach was seen from the technical and commercial environment, UNIX originally grew from a purely academic environment. The original weaknesses of UNIX derive from this very

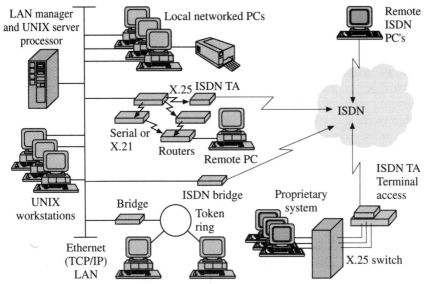

Fig. 5.3   A typical fully configured computer installation

fact. It was developed as a multiprogramming operating system with good communication facilities but it was considered weak on facilities to handle commercial files, had poor security and was designed to run on one processor. These early weaknesses, although now overcome, have left an impression of UNIX that has been slow to die and has resulted in a long acceptance cycle in the more traditional commercial environments. Within the engineering computing environment, however, UNIX is a very widely used operating system with the majority of CAD/CAM, CFD, FEA, CAPP and now EDM software running either on a UNIX workstation or server processor.

The general situation is now changing rapidly with the advantages of UNIX now being accepted more and more in the 'traditional' computing environments. For example, many UNIX-based warehousing, retail and manufacturing systems are now starting to rival sales of such solutions in proprietary environments. So while UNIX and the concept of open systems are closely linked (the term 'open systems' did not exist before UNIX came on the scene), open systems can be considered more of a philosophy under which UNIX, and other products of an 'open' nature, figure strongly.

This is certainly true of the EDM marketplace and products which we shall see later are, if not directly run under UNIX, certainly based on a DBMS which will run under UNIX. This is one of the main reasons for discussing UNIX and open systems at such length—any understanding of EDM requires at least a basic knowledge of the computer 'environment' on which the technology depends. It is hoped that this chapter provides this foundation.

5.4   Hardware considerations

In a similar manner to our understanding of EDM requiring a basic knowledge of open systems, we also need to consider some of the other elements of computer technology that relate specifically to EDM: the software and hardware elements.

This section will therefore outline some hardware elements. The software elements are discussed under Section 5.5. It should be pointed out that these subjects are massive in their own right and it is not within the scope of this book to describe them in any depth. The reader is referred to suitable IT-related reading material for such detailed information. However, EDM is a solution which can reach out to embrace some or all of these areas. They have therefore been included in order to clearly position EDM and increase our understanding of its technical requirements and functional scope.

Computing environments

The question is often asked, 'In what environment does a typical EDM system operate?' In truth there are no 'typical' EDM or engineering computer system environments. Just as each organization's markets, resources, strategies and cultures vary, so do the engineering computer environments and resources upon which the EDM (and other engineering software) will operate on. However, most companies now have:

- A 'corporate' computing resource, such as a mainframe or minicomputer with 'mission critical' data running commercial and manufacturing applications
- Departmental machines such as minicomputers and workstations (or PCs) to run departmental and local applications
- Personal desktop or portable machines to enable cost-effective and easy-to-use facilities such as word-processing, E-mail or spreadsheets on a local, as-required basis.

With regard to EDM in particular, the corporate mainframe will not in general initially be involved, apart from perhaps its use as a data/file server in some installations. This is because of their historical use in running large commercial and manufacturing applications where the processing is less complex but the data much more voluminous. In a mainframe-based manufacturing environment, for example, the often numerous terminals connected to such a machine may require to be 'ruggedized' to survive the adverse conditions usually found on a factory floor.

EDM systems are much more likely to be found in the minicomputer and workstation environments, and with UNIX-based systems in particular, as we have already discussed. This is because existing design and engineering applications which are closely aligned to the data that EDM will manage already

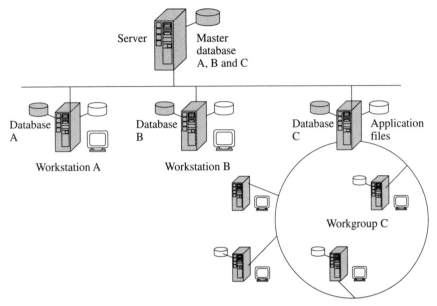

Fig. 5.4 An example EDM system configuration (*courtesy of CIMdata Inc.*)

run on such machines—facilities for interactive graphics and a high process-ing capability are two of the major elements on an engineering user's list of priorities. A minicomputer environment is one from where the main EDM software will operate, whether centrally or part of a distributed computing en-vironment. In this way, successful use can be made of the technologies we have described, such as database server machines, access to fast data storage across a network, print and image servers, etc. In order to clarify this sometimes con-fusing array of hardware, an example EDM system configuration is shown in Fig. 5.4.

One fact which must be mentioned here is the increasing use of PC technol-ogy. Workstations will be a common aspect of a typical EDM configuration but, as more powerful PCs are introduced and more engineering computer applications such as CAD are run on them, so the distinction between the two will become less clear. This trend is already well established and will continue as the cost of workstations declines and the power and corresponding cost of PCs increase. In general workstations are considered to have:

- More memory and faster CPUs
- Better networking capabilities
- CPU load sharing and parallel processing
- High resolution displays
- Faster graphics capabilities

Performance

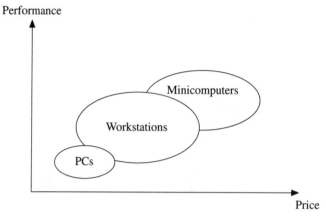

Price

Fig. 5.5 Price versus performance comparison

than PCs and the relative position of them is shown on the price versus performance comparison in Fig. 5.5.

Downsizing and rightsizing

The term 'downsizing' appears in many computer press articles and IT brochures, but what does it mean and how does it relate to EDM? In general, the term 'downsizing' is simply used to describe the practice of placing computer applications on smaller machines than they may have normally been placed on at a particular installation. These machines offer improved price/performance and lower running and maintenance costs than their predecessors. As can be seen this is a very simple and straightforward concept, but it can cause confusion due primarily to the term covering such a diverse range of different situations. It should also be noted that downsizing should be a business-driven process, not simply an IT-driven tactic considered in isolation. As a corollary to the term downsizing and perhaps to dispel some of the confusion, the term 'rightsizing' (and also 'smartsizing') was also coined and, as the name suggests, this means putting the application on the correct size of system—one which best suits the needs of the application. Although this may seem like an obvious thing to do, history bears testament to the fact that this is often not the case. Perhaps the terms should be used collectively 'Down/Right/Smart' sizing! In any event the trend is already set with mainframe systems being replaced with linked minicomputers and minicomputers being migrated towards PC networks with resultant, and in some cases significant, cost savings. See Fig. 5.6 for the expected downsizing trend over the coming years.

However, cost is not the only prime consideration. Rightsizing and the move to distributed computing means that systems can be developed that

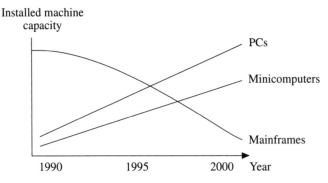

Fig. 5.6 Expected downsizing trend

allow local applications to be developed which also retain the integrated approach needed to meet business demands. In addition, it is possible in a distributed system to separate the data from the applications so that the information is available to all applications that need it, rather than have it directly linked to say an MRPII or accountancy package.

Processing speed

We are all aware that technological developments in the amount of processing power and speed which can be gained from a constantly reducing size of processor box is increasing at a dramatic rate. Any figures used to represent this would quickly become out of date but we can make a general comparison that any PC with a conservative specification of 2 Mb of random access memory (main memory or RAM) being driven at 15 Mhz (the processor speed) could be compared with a high-end mainframe of only ten years ago.

While it is important that any EDM installation should have adequate processing power, the need for high-end performance is not in the same league as that needed for graphics and CPU intensive computing applications such as CFD or 3D solids modelling. Workstation technology is predominantly used in these market sectors.

Storage media

EDM systems do require to have fast and reliable disk access to the stored data. Data in an EDM system can either be stored alongside the application software on a local disk or stored centrally on a network 'file server'. Whatever method of data storage is chosen, the following three main requirements must be met:

● High speed data storage and retrieval (with integrity)

- A cost effective solution
- High storage capacity

The main devices which will be considered here, apart from the primary RAM storage of the computer itself and the sequential access magnetic tape which is used for backup and general off-line data storage, are magnetic and optical disks. For convenience and efficiency reasons most EDM users, and other general computer users, will want to keep as much data on-line (i.e. directly accessible) as possible. Disk devices are important in that they provide this capability.

Data on (hard) magnetic disks (as opposed to floppy disks which are removable) is stored on invisible concentric tracks on the magnetized surface of the disk. The heads for reading and writing the data can move over any track to allow data transfer as the disk surface spins past. Modern hard disks tend to be small, sealed units capable of storing many gigabytes of data. Although hard disk drives are likely to be present in any EDM system, they are not generally used as the primary storage medium for images, due to the large volumes of storage required for such data. They are most often used for maintaining index files, acting as a buffer prior to transferring data to removable media, or as a temporary store for a period where documents are in constant use.

Optical disk technology is now becoming more accessible and affordable and can be considered under three main areas. Firstly, CD-ROMs. These are well established in the music business with the use of the compact disk. A pattern of recesses, which are pressed or burned on to the metallic coating of the disk surface, can be made to represent a sequence of digital data and this can be read using a laser beam within a suitable CD playback device. CD-ROM stands for Compact Disk, Read Only Memory and although the disk transfer rates are high and they can store large amounts of data, the storage is read-only because the CD itself can only be made by suitable specialist equipment. WORM devices (write once, read many) on the other hand can allow the user to write a suitable pattern to the disk itself for storage, but once written cannot be removed. This technology is ideal for backup purposes and can be of considerable use within an EDM environment in the archiving of released data, especially where audit trails are required. Finally, erasable optical disks are now becoming more widely used and act in a similar manner to traditional magnetic disks in that they can be written and read many times. They combine the advantages of this traditional storage mechanism with the fact that an optical disk is fast, can store vast quantities of data and are more rugged and less susceptible to contamination by magnetic sources or dust particles.

Other hardware elements typically found in an EDM environment include:

- *Electrostatic plotters*   These are much faster (up to 20 times) than the more traditional pen plotter, are high precision machines and can be left running unattended overnight. They are, however, more expensive and require to be justified correctly if the maximum return on their investment is to be made.

- *Scanners*   These provide the ability to scan up to A0 drawings and documents into raster images for storage and display purposes. Intelligent raster-to-vector convertors are also available to enable the fast and accurate generation of CAD-ready files.

5.5   Computer software

In Fig. 4.1 we saw the different 'levels' or 'layers' of computing technology which separate the user at the top and the hardware at the bottom, and have already looked at the database and hardware layers in some detail. This section will concentrate more on operating systems and computer software technology in order to provide the necessary knowledge base to understand IT in general and EDM in particular.

Proprietary operating systems

In simple terms an operating system is a program which controls the operation of the computer itself. The main part of it (the operational kernel) usually resides in the computer's memory and this controls the flow of data from the application program to peripherals such as disks and printers in addition to managing the other computer resources such as memory allocation, processor time and arithmetic operations. The architecture of the operating system itself is related to the computer on which it operates and this initially led to operating systems which were specifically suited to each manufacturer's machine or range of machines. These are known as proprietary operating systems. The more widely known include:

- *VM and MVS from IBM*   These two operating systems are mainly used for large-scale data processing applications as found on commercial mainframe and minicomputers.
- *VMS from DEC*   One of the first successful and truly scaleable operating systems which runs on the VAX range of computers. It was introduced in 1978 and has become the reference model for 32 bit minicomputer systems. VMS provides an extremely powerful environment for developing and executing mainly technical and departmental computing applications, but can also be found in a wide number of commercial environments.
- *DOS from Microsoft*   This has been developed and marketed by Microsoft as a single-user operating system; the first version was released in 1979. Due primarily to the fact that IBM adopted it for their PC which was introduced in 1983, it has become the *de facto* standard for personal computing and is the most widely used operating system in the world with an extremely wide range of applications.

There are many other proprietary operating systems on the market, each having their own range of applications. They are generally incompatible with each other but usually perform their intended functions well.

UNIX and X-Windows

The growth in the popularity of UNIX is closely aligned to the developments in processor technology. With improvements in power, capacity and price, it is now feasible and cost effective to install UNIX machines into a networked business computing environment. In addition X-Windows, a machine-independent window manager, has been developed to ease the handling of graphics requirements for the newer range of high-performance workstations running UNIX. UNIX is a complete and well-established operating system that allows many users to run multiple applications all at the same time, sharing resources and information. It can be implemented across numerous machines, including IBM, Bull, DEC, Unisys, Olivetti, ICL, HP and Sun.

UNIX was developed at AT&T's Bell Laboratories in 1969, has been developed ever since that time, and is now virtually universally available. Workstation networks are now replacing older timesharing multi-user systems, which utilized many 'dumb' terminals connected to a central computing resource, be it UNIX or proprietary. Today's workstations are sophisticated pieces of equipment, with high-resolution colour graphics displays and windowing software specifically designed to operate on a network. The network's central computer, or file server, could also be UNIX-based with the workstations connected to it through a networking standard known as Ethernet, using the TCP/IP networking protocol (as explained in more detail in the next section of this chapter).

As we have already mentioned, early UNIX gained a reputation for being an unfriendly system—it had, after all, been developed by academics who were very happy with its rather cryptic user interface. As its use spread, so too did the need for a more usable interface. New users looked for facilities like windowing to let them run several applications on their screens at the same time, moving from one window to another as they wished. They also wanted an attractive user-friendly graphic display, so that they could operate the system by pointing and clicking on icons with a mouse. The X-Windows system, which was designed by the Massachusetts Institute of Technology (MIT) met these requirements and has now become the *de facto* graphics windowing system for UNIX. X-Windows is a network transparent presentation manager which means that it is not restricted to any specific type of network but can run on any machine which is attached to the network.

The window managers themselves, i.e. the applications which provide the 'look' and 'feel', actually define how the interface appears to the user. Although there are a number of these on the market, the two main players are

MOTIF from the Open Software Foundation and OpenLook from Sun/AT&T. We can see therefore that UNIX, X-Windows and TCP/IP are key components to open systems installations.

Application development languages
Certain third generation languages are briefly outlined here and lead on to the subject of fourth generation languages (4GLs).

FORTRAN
Fortran (which stands for FORmula TRANslation) was developed by a research group within IBM in the mid 1950s to enable systems to be written without using assembly or machine code directly (as had been required up until that point). It was defined under an ANSI standard in 1966. Fortran was originally designed for scientific and engineering computing applications and many widely-known packages are written in this language. The original short-comings in the language were recognized and a major new release issued in 1977, rather predictably known as Fortran 77. Many other additions have since been made, some standard and some of a 'proprietary' nature by, for example, IBM and DEC.

ADA
An extensive survey by the US Government's Department of Defense (DoD) in 1973 showed an annual DoD spend in the order of $3 billion on software that was being written in no less than 1500 different language variants. Various initiatives followed in the search for one language to replace this undesirable situation and, by 1979, a final selection had been made from a European consortium, led by the then Honeywell-Bull. So ADA was born and adopted as an ANSI standard in 1983. It is based to some extent on Pascal but with a number of additions and is considered to be a highly complex but reliable language.

C
The C programming language is closely associated with the UNIX operating system since it itself is almost entirely written in C. It was developed as a low-level language which would support the efficient coding of performance-critical systems, such as operating systems, and grew out of two ancestors, BCPL and B, which was originally developed by Ken Thompson in 1970. The accepted definition of C was contained in the original book by Kernighan and Ritchie but is now an ANSI standard. The language has undergone many

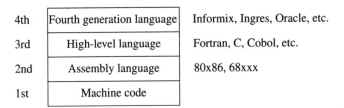

4th	Fourth generation language	Informix, Ingres, Oracle, etc.
3rd	High-level language	Fortran, C, Cobol, etc.
2nd	Assembly language	80x86, 68xxx
1st	Machine code	

Fig. 5.7 The historical development of programming language levels

improvements since its inception and indeed C++ has become one of the major 'object oriented' programming languages which are becoming more widely used in today's complex leading-edge application programs.

4GLs
The term 'fourth generation language' was initially coined by James Martin Associates and was used to describe the concept of a language at a higher level than that previously used. The levels of programming language are shown diagrammatically in Fig. 5.7.

However, the term is now generally accepted to mean high-level languages and utilities associated with relational database management systems, although it is very difficult to find a simple, easy-to-understand definition of what one actually is. Defining a 4GL is a bit like describing an elephant—you can recognize one when you see it but it takes longer to describe verbally. In general, a 4GL is considered to be a complete, integrated, interactive development and 'run-time' environment which includes:

- file/database handling
- independent report definition
- independent screen definition tools
- individual database operations

These are often presented in two ways: one relating to the programmer's development language, and the other to a user's tailoring language. 4GLs enable the rapid development of applications through 'simple' interfaces and high-level utilities and have proved to be major productivity aids in a wide variety of software development environments. Their lack of standardization and definition does, however, emphasize their proprietary nature although many 4GLs will now operate in conjunction with a number of different RDBMS systems.

5.6 Networking and communications developments
Local Area Networks (LANs) and Wide Area Networks (WANs) have evolved from traditional methods of networking and communications. Their

development has been, and will continue to be, driven by the need for information exchange (and the nature of this interaction) in local work environments. Traditional workplace developments closely parallel the evolution of distributed processing in that, just as interdepartmental communications between distributed processors became essential during the 1970s, so the need to connect systems within and between work environments is essential now.

A typical modern workplace contains many intelligent machines to assist in the everyday activities and communications we now take for granted, such as PCs, copiers, plotters, facsimile machines and printers. In order to allow functional groups of people to make the maximum use of such equipment, and to function in an efficient and integrated manner, these machines need to be able to communicate and exchange information quickly, easily and reliably. The associated computing environment should therefore successfully provide interoperability such that tools and databases can communicate as part of a common network.

LANs are particularly important to allow the required access to information to be achieved within a business organization. Various studies have shown that around 80 per cent of the information used in a local environment is actually generated from within the environment, and of this information a further 80 per cent will be used solely in that local environment and only 20 per cent used outside.

WAN implementations can encompass a broad range of solutions, and are usually considered in three main classes. In increasing cost order they are:

- PSTN (public switched telephone network) systems
- Leased line systems such as Kilostream or Megastream
- Wireless systems using radio broadcasting or line-of-site lasers

There are two main elements to be considered in any networking environment: the physical communications aspects such as cabling and ducting, connections and the increasing use now being made of fibre optics etc.; and the actual communication protocols themselves which run on these physical layouts. These aspects are covered by numerous standards, some of which are:

- *Ethernet*   In 1976 Rank Xerox published the description of a coaxial cable network in which the nodes monitor the cable (the Ether) during transmission. If a data 'collision' was detected then the transmission was terminated and restarted. This was known as Carrier Sense Multiple Access with Collision Detect (CSMA/CD), or *Ethernet*, under its trade name. Xerox, DEC and Intel published a specification for Ethernet in 1980 and it has since become an ISO standard. Ethernet is a simple LAN system based on coaxial cable using a 10 Mbps (megabits per second) transfer rate and is flexible enough to link many devices together and provide a high-speed file transfer mechanism.

- *X.25* This is a set of protocols defined by the Comité Consultatif Internationale de Télégraphique et Téléphonique (CCITT) in 1976 to enable a common standard to be established between countries with regard to public network services. A public network service is similar to or part of the public telephone service, where a private company will offer network services to any organization which wishes to subscribe to it. X.25 uses a technique known as 'packet switching' which is usually considered uneconomic for small networks and is therefore to be found mainly in WAN configurations.
- *X.400* The CCITT X.400 committee was formed in 1980 and X.400 was issued as a standard in 1984. Its operation is analogous to the way a post office works, in that messages are posted by a 'user agent', and sent to their final destination by 'message transfer agents'.
- *TCP/IP* TCP/IP stands for Transmission Control Protocol/Internet Protocol and was originally defined by the US DoD. Although the TCP/IP protocols were defined before the OSI model was developed, they do generally conform to it. TCP/IP can be defined as a set of processes which allow two or more computers to communicate together and can be distinguished from other network protocols such as X.25 in that it includes an outline definition of all levels of the interface. TCP/IP is well suited to LANs, and in particular to those using Ethernet. It is simple to use and is now the *de facto* standard for linking UNIX systems together. Its success in this area may well be due to its vendor independence.

5.7 The effect of standards

The word 'standard' has been used many times in this chapter, just as many of us use it in our everyday working lives—but what standards are we referring to and how have they evolved? Who defines and controls them? These are some of the questions we will consider in this section. We shall begin by considering those which specifically relate to open systems and follow this by those of a more generic nature.

Open system standards

Standards-making bodies develop and agree specifications for open systems and, in general, it is left to the IT industry to interpret these and turn them into working systems. Although standardization in IT has been talked about for many years, it is still an evolving concept, but there are enough established standards to let users start to implement open systems in any area of business. Before briefly considering the standards themselves, it is probably worth distinguishing between the three main types of standards bodies which directly or indirectly relate to the definition of open systems given in Section 5.2. Firstly, there is the international or national standards organization such as the ISO or British Standards Institution (BSI). Secondly, there are supplier alliances who may decide on some standards as being beneficial to their common interest, such as the Open Software Foundation (OSF). The third type is not so much a standards body but relates to those suppliers who have achieved such a high market share as to create *de facto* standards such as Microsoft's PC operating system MS-DOS.

The OSI standards previously referred to have been formulated to eliminate the barriers to global communication and several bodies have collaborated to formulate communications standards. These include the Institute of Electrical and Electronics Engineers (IEEE), the International Standards Organization (ISO) and the CCITT. CCITT is primarily concerned with data communications and telephone systems and its recommendations for the interface between a computer and a packet-switched network has now become an international standard (X.25). OSI is based on a seven-layer model for network architecture, known as the Open Systems Interconnection Reference Model. The purpose of each layer is to offer certain services to the higher layer without detailing how those services are actually implemented. Users get the impression of communicating directly with another machine or process while the network layers insulate them from needing to know what has happened to the data or its passage through the network. This is an essential feature of open systems and is known as transparent communication.

Before OSI standardization, communication protocols (conventions that set up and maintain data communication) were many and varied. For example, TCP/IP predates the OSI model but its use is so widespread that, as we saw in the previous section, it has become a *de facto* standard. Others are standards—such as X.25. They are supported by Ethernet and IBM's Token Ring (both IEEE 802 standards) which are widely used in LANs.

Portable software, i.e. that which can be used over a number of different machines and operating systems, demands a common operating environment or platform. This platform, which consists of a computer and its operating system, supports the application programs themselves, such as financial ledgers or word-processing. Standards in portable software have developed along *de facto* lines, i.e. by general acceptance and use, rather than by dictate (*de jure*). One of the most widely used single-user operating systems is Microsoft's DOS, now standard on most desktop computers. Although it will only work on the Intel family of microprocessors, there are so many copies of MS-DOS in use worldwide that it does provide a common operating environment.

Conversely, the UNIX operating system has been designed for multi-user, multi-tasking operations and has benefited from over twenty years of development and refinement. It is now one of the most portable operating systems in the world and runs on all types of hardware, from a PC to a Cray supercomputer. It has been adopted by all the major hardware manufacturers and

UNIX systems based on Reduced Instruction Set Computers (RISC) are gaining ground, especially in mid-range, high-performance configurations. Portable industry standard protocols facilitate the transfer of files. This means moving and accessing not just single files but complete file systems from one machine to another. The Network File System (NFS) is found on almost every UNIX network, and has become an industry standard due to its public availability and user and protocol transparent nature.

Both MS-DOS and UNIX provide a firm basis for long-term expansion and growth. They provide a common operating environment which benefits both software developers and users alike. Some of the formal international standards organizations which contribute to the specification and acceptance of many widely known standards include:

- The International Standards Organization (ISO)—a United Nations agency, which acts as the umbrella organization to which the various national standards-setting bodies are affiliated
- International Electrotechnical Commission (IEC)
- International Telecommunications Union (ITU)—which is fed by the CCITT

Formal national standards organizations within the UK include:

- BSI—the official UK body on the ISO
- National Computing Centre (NCC)
- Department of Trade and Industry: IT Standards Unit
- The Institution of Electrical Engineers (IEE)
- British Computer Society (BCS)

Formal national standards organizations within the USA include:

- American National Standards Institute (ANSI)
- The Association for Computing Machinery (ACM)—US equivalent to the BCS
- US Department of Defense (DoD)—although not part of the usual standards setup, they do create (and enforce through contract terms) many standards, recognized by the general form MIL-STD-*nnnn*.
- The Institute of Electrical and Electronics Engineers (IEEE)
- National Institute of Standards and Technology (NIST) (formerly known as the National Bureau of Standards)—NIST is not a standards-setting body exactly: it acts more as a calibration/accreditation/conformance testing body, and also performs standards-related research

A description of the supplier alliance organizations and a number of the relevant major standards themselves are included in the Glossary (Appendix B). It should also be mentioned that the national standards organizations in Europe include:

- AFNOR (Association Française de Normalization)—the French national standards setting body
- DIN (Deutsche Institut für Normung)—the German national standards setting body

Other standards

There are numerous standards which could either be said to relate directly to EDM, or at least have some effect on its use, related procedures or associated areas of technology. The main ones are briefly explained below.

Quality standards

These include:

- *BS 5750*   The British Standard for Quality Systems. ISO 9000 is the direct international standard equivalent to BS 5750. (EN 29000 being the European version). It is split into a number of sections:

Part 0:	Principal concepts and applications
Part 1:	Specification for design/development, production, installation and servicing
Part 2:	Specification for production and installation
Part 3:	Specification for final inspection and test
Part 4:	Guide to the use of part 1

    with others in preparation for future release.
- *AQAP 1*   AQAP stands for Allied Quality Assurance Publication. This was originally defined as a stringent Ministry of Defence (MoD) marine standard for the control of design and manufacturing system procedures (AQAP 13 relates to software development). AQAP standards will now gradually give way to ISO 9000.

Worldwide data exchange formats

The most widely known of these is the CALS standard. CALS stands for Continuous Acquisition and Life-cycle Support (previously Computer-aided Acquisition and Logistics Support). It is a strategy designed by the United States Department of Defense (DoD) in cooperation with industry as a solution to the increasingly complex problem of procurement and the need to exchange agreed levels of technical information between many parties, by moving from the current paper-intensive method of working to a more digitally-based environment during the 1990s. It aims to replace paper transfers between the DoD and its suppliers by digital transfers, in areas such as product data, support data and technical manuals. This will be achieved in two phases:

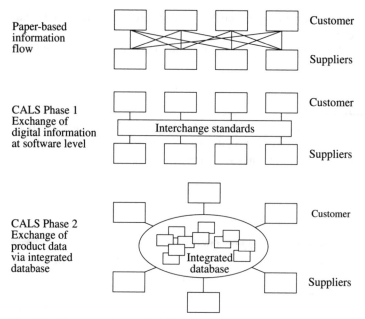

Fig. 5.8   The evolution of CALS

- In Phase 1 a set of standards and specifications is being used for the transfer of digital information via magnetic media from prime contractors to the Government.
- In Phase 2 this information will form large databases which will be capable of being accessed over standard communication networks.

These are shown in Fig. 5.8.

Phase 1 is in operation now, with many contractors and subcontractors (of both a defence-related and commercial nature) reaping substantial rewards and preparing for Phase 2. The benefits to be gained from using CALS include:

- Obtaining the right information at the right time
- Improved international competitiveness
- Better third-party relationships
- The ability to keep pace with your customers
- Improved 'internetworking' with other companies, engineers and key functions within the organization.

For example, the US Navy's Automated Logistics Publishing System (ALPS), which is an electronic publishing system designed to automate the creation and revision of technical manuals, saved $1 600 000 in one year with a $400 000 investment. The system cut costs by 70 per cent and reduced worktime by 80 per cent in less than six months. In one case, a job that was expected to take six months took six days!

CALS was first published as US DoD MIL-STD-1840 in September 1986 and was updated in December 1987 to MIL-STD-1840A, and although it has grown from a defence background, its use is expected (like many previous standards) to spread into the commercial and private sector business operations. Strong evidence of this can already be seen. CALS is more of a compendium of standards than a standard in its own right—it lists a set of existing (or draft) standards, or defined subsets of them, for data exchange, and adds an overall umbrella format within which these exist (e.g. covering tape formats). The base standards used in Phase 1 are:

- Standard Generalized Markup Language (SGML), also known as MIL-M-28001. This defines an international standard used for technical publications which can also contain graphics and scanned images. It enables more structured documents and information sets to exist. An application of SGML also exists and is used to access information interactively including sound and video sources—a multimedia approach for what are termed interactive electronic technical manuals (IETMs).
- Initial Graphics Exchange Specification (IGES) subset known as MIL-M-28000. IGES is a standard developed to enable the exchange of product data of a 2D and 3D graphics nature between various CAD and graphics systems.
- Computer Graphics Metafile (CGM) which is a simple format for representing 2D illustrations with geometric graphics objects as found in technical illustrations. This was originally defined using ANSI X3.122.
- CCITT group 4 raster, a CCITT recommendation for representing and sending scanned images. Often used for engineering drawings and illustrations.
- Electronic Design Interface Format (EDIF) to allow electronic design data to be transmitted.
- EDIFACT—EDIF for administration, commerce and transport. This allows commercial data, such as quality information, orders and payments, to be exchanged worldwide.

CALS will develop in line with the work of other standards, so it is unlikely ever to be a static standard. For example, the IGES standard will eventually give way to PDES/STEP. IGES is primarily used for exchanging data between different CAD/CAM systems: PDES/STEP extends this concept and provides a complete and unambiguous computer definition of the physical and functional characteristics of a product.

In summary then CALS is about integration, not just standards and specifications. Its aims are to improve the timeliness, reduce the cost and improve the quality of manufactured products and their supporting technical data.

Achieving such objectives will lead to improved operational performance and industrial competitiveness. CALS includes and facilitates:

- Integrated logistics support (ILS)
- Concurrent engineering practice—to such a degree that CALS is often termed CALS/CE
- Computer-integrated manufacturing (CIM)

Functional standards
These include:

- *DEF.STAN. 05-10* A UK Ministry of Defence (MoD) standard (DEF.STAN.) which specifies the MoD requirements for the preparation of engineering drawings and associated documents. It is applied in conjunction with BS 308 and BS 5536.
- *DEF.STAN. 05-53* An MoD standard relating to the identification of drawings and their retrieval.
- *DEF.STAN. 05-123* This relates to the control of designs and design records.
- *DEF.STAN. 05-57* The MoD standard for configuration management.
- *BS 308* Engineering drawing practice.
- *BS 4732* Magnetic tape labelling and file structure for data interchange.
- *BS 5536* A specification for preparation of technical drawings and diagrams for microfilming.
- *BS 7000* A guide to managing product design.

Other standards
These include:

1 *PHIGS—Programmers Hierarchical Interface to Graphics Systems* This is a graphics standard which is intended to provide applications portability between displays (or windows if used with X-Windows) from different vendors.
2 *PDDI—Product Definition Data Interchange* The purpose of this standard was to provide a mechanism for product data interchange that overcame some of the deficiencies of IGES by addressing manufacturing information such as datums, tolerances, machining technology, etc.
3 *VDA-FS—Verband Der Automobilindustrie-Flachenschnittstelle* VDA-FS was developed following the adoption of IGES within the German automotive industry, in recognition that IGES was inadequate for handling sculptured surfaces between CAD systems. (DIN Std. 66301.)
4 *SET—Standard D'Echange et de Transfert* This AFNOR standard was designed to provide a more complete exchange specification than IGES, for

both interchange and archival, and to be more compact than IGES. Used extensively by the aerospace industry.

5 *DMIS—Dimensional Measuring Interface Specification*   Designed to provide a two-way exchange of data between CAD/CAM systems and computer inspection equipment (CMMs, optical gauges, etc.).

6 *DXF—Data Exchange Format*   Originally a proprietary exchange format from Autodesk Inc. (now a *de facto* standard) to provide a means of easily exchanging data between AutoCAD systems. Now widely available from all CAD vendors.

The above is only intended as a brief guide to some of the major standards which may influence CE/EDM/CIM—there are many, many more. It is not within the scope of this book to detail them all: the reader is referred to the standards bodies mentioned in the text as sources of further information. The effect of standards on currently accepted computer technology has been significant. They are now an established and growing part of today's computing environments, but it has to be said that in some ways their proliferation can also cause problems and confusion to many users and developers alike. As some wag was heard to say recently, 'The good thing about standards is that there are so many to choose from.' It is usually best to get advice on those which you are unsure about and which affect your own operating environment.

6

Justifying EDM

6.1 Why invest?

The philosophy of the engineering and manufacturing industry has changed dramatically in the last thirty years. The generally accepted major hurdles confronting future success relate to the acceptance of continuous change and making the maximum competitive use of time. Engineering-related IT is an essential foundation for achieving these goals in today's business environments and is required in order to maintain both steady improvements in manufacturing efficiency and the quantum improvements in performance we can now begin to appreciate.

Investment in some form of information technology is no longer considered a luxury—a continuous programme of improvements in IT and related business processes and philosophies is mandatory, with most engineering and related companies investing in some form of computer assistance, even if only a humble PC. Other manufacturing concerns have spent extremely large volumes of cash on certain projects with sometimes very questionable returns on the investment. Why? How can we ensure that investing in EDM does not end up in this category? These are some of the questions we will be considering in this chapter. Before continuing it is advisable to reiterate the fact that any investment strategy, whether relating to IT or otherwise, must be developed in line with the corporate business strategy. This simple point is one of the main reasons why many investments fail: the IT strategy must reflect and support the organization's long-term aims. For example, one large engineering company was recently commended for admitting a mistake which occurred in the early 1980s where, in their initial pursuit of engineering excellence, the company spent a large sum of money on a comprehensive CAD/CAM system. This increased their outer body-shell capacity to 42 units per month. The company realized later that the highest added-value in the product concerned was not in the outer shells, which almost any engineering company could make, but in the more specialized internal mechanisms involved and their associated electronics. Unfortunately, the company had not invested in these areas. Worse still, in the entire year during which the company had spent all its budget increasing its capacity to 42 units a month, they only received orders for a maximum of 34.

A further example comes from Professor Colin New of Cranfield School of Management, who cites one company in the food business which built a £5m glass jar bottling line. Six months before the product launch, marketing announced that the product would not be packaged in a glass jar, but in a composite container! Furthermore, the jar line could not be turned over to something else since up-front budget limitations had restricted the line of jars to a certain size and below. The company concerned did not make any other products in that size range. 'If they'd spent another £400 000 on the £5m they spent anyway,' said Professor New, 'they could have had a multi-size jar line which could have packed the other products made in the plant.'

A final example, also from Professor New, relates to a chemicals company which wanted a plastics plant whose remit was to be the lowest unit cost manufacturing plant of its type. Three years and £600m later, manufacturing took delivery. The designers ensured low unit costs by raising the yield as high as possible. This meant using a 27-stage continuous process which held the product for 60 days. Then marketing told manufacturing they had to make 70 different product variants, so if someone ordered grade 70, they would have to wait almost 12 years for delivery! The company concerned ultimately had to choose between accepting a restricted product range or a 27 single-batch stage process. This has led to it being the highest unit cost plant of its type in the world.

The common thread running through all of these examples is that large sums of money could have been saved if only different departments communicated effectively. If this communication and a process of continuous improvement is not made, the leap required to catch up with world-class companies can sometimes be too great—certainly errors of such magnitude as the third example can only be sustained by the very largest of companies. Of course, investment is difficult to maintain on an ongoing basis, and equally difficult to justify. Not only does it depend on management foresight and commitment, sufficient allocation of funds and budgets, but also on more global factors such as government policies and grant aid. With hindsight we can appreciate that, at least in the UK, the early 1980s were a golden age for IT and Government. Ministers for Information Technology were appointed and funding initiatives established, most of which have now sadly disappeared. An investment infrastructure such as this now no longer seems to be such a Government-centred responsibility—it is therefore up to each business enterprise to plan and secure its own future. The problems experienced with such a technology-focused approach are now well known, but certain investment and financial aid initiatives are always required to ensure that we invest for the future rather than worry about it.

The aspects of traditional accounting models also require to be considered in today's turbulent manufacturing and trading environments. It is now widely realized that the previous belief that service industries alone could sustain any form of significant economy is now no longer true and that manufacturing industry itself has to ensure its own future. The days of blaming the Government, management, unions, our respective engrained cultures or other influences are over. In line with this belief, we need to re-evaluate certain of our more established mechanisms for investing in new technology. For example, ten or twenty years ago when business life was fairly stable, an analysis of past performance was seen as providing a sound basis for future decisions. Accountants were trained to identify the truth about the past by answering questions such as, 'How well did we do what we chose to do?' with the emphasis placed firmly on tactical management and achieving the budgeted profit for the year. Today's business environment is different—change is happening so fast that this situation is no longer true and, in such a climate, profit is not the only valid indicator of financial health. For example, mismanagement of cash flow can cause large, successful and well-managed companies to crash, just as easily as smaller enterprises. Also, many companies which focus on short-term results (such as measuring performance and setting insular objectives) have lost market share to competitors from countries such as Germany and Japan where accounting numbers are not held in such awe. And so in today's situation of rapid change, strategic management predominates and other questions such as, 'Was our choice of what to do correct?' and 'What should we do next?' need to be answered. These types of questions, as we have previously discussed, cannot be answered purely in traditional 'accounting' terms and require a broader, more strategic 'financial and management' perspective to be taken—an aspect which is being accepted by an increasing number of leading companies and business enterprises who have accepted the concepts of business excellence.

6.2 Project implementation

Like many projects, the implementation of an EDM system can be viewed in a number of ways. In the simplest of scenarios, an off-the-shelf product can be purchased which, together with the appropriate computer hardware, can be installed on-site and after suitable training the users are told to get on with it. This approach would not require a significant implementation strategy, the projects would be considered along traditional 'commodity' lines, and the ultimate benefits would definitely be questionable in all but the simplest of installations.

In order to gain the maximum advantage from the adoption of EDM, we should consider it in a more global sense as we have already discussed thus far. We need to use it to question traditional practices, to use it in new and enterprising ways, to help bridge and integrate perhaps existing but discrete processes, etc.—all aspects which extend the above basic scenario into one which can involve many company departments and working procedures. When considered in such a manner, the implementation of an EDM system extends beyond simple hardware and software elements and requires a more comprehensive implementation schedule to be adopted.

EDM technology cannot be used to 'paper over the cracks' in an existing engineering environment. If a company's existing engineering practices show signs of weakness, simply adopting EDM will not be likely to cure them. The root problems require to be identified and addressed prior to the adoption of EDM—a fact very often omitted from many system plans, proposals and justifications. What is needed here is innovation, not automation of what exists.

The existing experience gained from other investments in engineering-related information technology (such as MRPII) can prove to be of great benefit. By re-using these existing and generally well established procedures, the implementation of a successful EDM system should prove to be less of a 'pioneering' type of project.

Because any project of this type cannot be considered on a 'single-shot' basis and, like computer-integrated manufacturing itself, EDM operates over multiple divisions within an organization (and is a unifying technology), it is very difficult to see where to start while maintaining an overall view of where you want to get to eventually. The implementation aspects of EDM will be considered in more detail in the next chapter, but before we continue with our justification of EDM here, we need to correctly position a project of this nature. The types of problems that EDM can help address, such as file management and change control processes, also require strong teamwork and cooperation to exist within the company concerned. If these qualities either do not exist or are not actively being fostered from an *operational* viewpoint, then any *system* which attempts to bridge them will ultimately fail. The implementation of an EDM system can certainly be used as the catalyst to get such cooperation moving, but there needs to be a general willingness to do so and an understanding of the benefits to be gained.

This is similar to the actual implementation of the system itself, where a teamwork approach and an understanding and commitment is required if a successful implementation is to be achieved. An education process is required to address the 'We've tried this all before', 'It'll never work' or the 'The computer systems here are all rubbish anyway' type of people who will have a negative effect on any form of major project implementation.

One further point to bear in mind is that implementing EDM generally takes much longer than even the most conservative initial estimates would have us believe. This has to be borne in mind when initial timescale estimates and costs are established—there are generally many unforeseen procedures and problems which arise in an implementation of a solution of this type, especially in a more established, traditional organization where change is not a readily accepted idea.

Finally, as with any IT project of this nature, a phased approach needs to be established with suitable milestone events, a quality focus, risk management, etc. defined according to the individual EDM system chosen and the environment under consideration. Phases could include requirements analysis, tender preparation and evaluation, contract negotiations, solution development and integration, implementation planning, pilot implementations, system acceptance and so on.

6.3 EDM project fundamentals

The implementation of a successful EDM system not only requires a detailed justification to be put together in terms of the technology itself with the associated hardware and software elements, but it also relies on certain baseline decisions being documented and discussed to provide a secure foundation for the other aspects to build on. Such elements are not unique to EDM system implementation—they are generic in nature and are mentioned in overview form here since many of them may already be well known to readers. Many of them are also difficult to cost. We have already considered these intangibles under the benefits sections of Chapter 3, but mention of them is also made here since they will need to be included in any EDM justification. Some of the points noted below will be considered in more detail in the next chapter.

Quality

The aspects of quality within this context can be considered in two main areas, both of which require to be analysed when justifying EDM. These are:

1 Quality of implementation. Specifically:
   (a) a quality plan related to the implementation schedule for the solution in question
   (b) the quality of the solution itself in terms of hardware, software and services as supplied by the chosen vendor
   (c) the quality of the team and persons involved in the implementation.
2 The impact on the quality aspects of the organization itself as a direct result of implementing EDM—i.e. the Total Quality aspects—for example the ability to provide better products, faster service or improved information. This is especially important if the company concerned is already working in a TQM environment, but equally may be used by any other company to initiate such a climate.

Funding

The factors relating to how a project of this nature could be funded will generally be closely tied to the individual company's operating procedures and financial situation. In a well-managed environment, budgets for the installation and implementation of EDM may already have been established as part of a longer-term plan. But equally, and probably more likely, the decision to proceed with EDM may be taken in a more uncoordinated way, perhaps sparked off by a serious product recall situation or as a result of impulse management decisions after suitable education and awareness.

Also, the costs of EDM, because of the generally accepted need for a higher consultative and service element than say perhaps CAD or DTP, could be split between capital and expense budgets. The capital budget could be used to purchase the initial hardware and software elements, with consultancy, implementation and training taken as required from expense budgets on an ongoing basis. Even the initial capital budgets could be split between the different departments concerned. For example, if a modular solution was purchased, the DP department could pay for the hardware and service elements, while each department paid for the relevant parts of the solution to enable them to perform their required function.

Structure

Any EDM project plan must detail and justify the resources required to ensure a successful implementation. One of the most important resources to consider is manpower. This is an often overlooked or underestimated resource, since traditionally hardware and software elements have taken the lion's share of any IT project of a similar nature to EDM. Information relating to the types, numbers, grades and structure of the staff required in the implementation process therefore requires to be detailed, costed and justified to senior management.

Requirements

'Surely this is obvious,' I hear you say. 'How can anyone look for a system without first defining what they want?' Well, surprisingly enough, many people do. This may be generally accepted during the initial stages of any system purchase where an introduction to the technology is required, perhaps at a tradeshow or roadshow type event. However, beyond this stage a list of requirements must be generated to focus often diverse ideas on needs and requirements, and to ultimately form a major element of the tender document, thus ensuring the best solution is chosen over both the short and longer term.

Solution

Choosing a system is not as easy as one would imagine. There are many aspects to be considered, some generic and some specific, and as mentioned above the choice is even more difficult without a set of predefined requirements. Consideration should be given to whether a customized or standard software solution is required and to what degree the system should be tailored. If it is tailored, should the internal DP/IT department undertake the task, or should it be an external software developer or the system supplier?

External help

EDM system suppliers will generally be keen to provide access to their implementation consultants, but a decision is required as to how far this is used and in what specific areas it is most required. Consultancy skills and external assistance can also come from the well-known international consultancies (PA Consulting, KPMG, Ernst & Young, etc.), from systems integration specialists (such as Bull Information Systems, DEC, ICL, etc.) or from specialist EDM consultants (such as CIMdata). In general, EDM is not something which will be well known in an organization: it is a whole new environment for many. Companies installing it very often need access to people who have experienced it before. An external consultant can provide this objective view and help of this nature can prove to be a significant factor in implementing a successful system—it is an unfortunate fact of life that the first item to be taken off a quote which may be too high is consultancy (be it in a pre- or post-installation form).

Project management

Of all the projects undertaken by organizations—from total corporate restructuring to the simple introduction of a new product line—it is the IT project which most often suffers delays. All too often, IT projects run well over budget, fail to deliver the advantages they were designed to introduce or run months, and in some cases years, late with extremely damaging results. Why? Very often it is simply because by their very nature they are new to the organizations undertaking them, and correct control over them is not established or maintained. As a result of this many new methodologies have been established over recent years to assist in the management of IT projects such as EDM and thus minimize the number of such 'runaway' projects. Correct project management is therefore of paramount importance, particularly in the areas of risk management, milestone events, planning and paperwork.

Reporting

Reporting and communication are vital aspects to consider in any project and are considered further in Chapter 7.

Problems

Problems? What problems? Surely there won't be any if we manage the project correctly? Many people assume this to be the case but unfortunately it is not so. Accepting that there will be problems is half the battle—once this is understood then plans can be established to anticipate them, to plan, categorize, assess, value and make contingency plans for them. By doing so they can be managed correctly—again part of the project management function—and a successful conclusion to the EDM project achieved (if ever such a point is reached—there are many aspects which can be added to the concept of EDM as technology, understanding and imaginative use of it grows).

Non-financial considerations

These are the benefits which can be gained from the successful use of EDM but which are difficult, or impossible, to cost. They include the 'opportunity' cost of having a successful EDM system, perhaps as a result of obtaining business which would otherwise not have been gained, or of having the correct data on which to base business decisions where before there was none or only distrust in the computerized information made available. EDM, as part of an overall IT strategy, also needs to be considered as part of the organization's strategic business plans over at least a 3 to 5 year period. One final project fundamental can be adequately demonstrated by the saying, 'Only apply the lotion where the wound hurts'. In other words, only use EDM where there is a direct benefit to be gained, rather than applying the technology purely for its own sake, or blindly adopting it simply because everyone else seems to be doing so.

Many people find that, after the initial scope, target areas and objectives of a project such as EDM are initially defined, the project undergoes continuous expansion (if not correctly managed!) into those areas where the benefits are less desirable or obvious, with resulting dilution of effort and enthusiasm for those areas originally identified. This 'scope creep' as it is sometimes known can also arise from the regular movement of the project 'goal posts' and changed objectives, as well as constant expansion of the project. All of these factors result in the project attempting to be 'all things to all men' with the result that milestones are not met, key benefits are not obtained, enthusiasm is lost and the project begins to spiral downwards into the 'runaway' type we have previously described. Key elements must always be borne in mind and the project defined in such a way that the energy of the project team is constantly focused on beneficial objectives as shown in Fig. 6.1.

6.4 Financial and costing considerations

Having now achieved a greater understanding of EDM as part of a concurrent engineering environment, looking at its constituent elements and how they

The project definition exercise ensures that the energy present in the project team is focused on the beneficial objectives

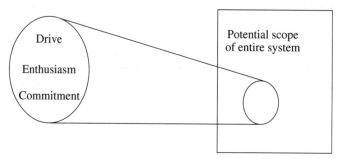

Fig. 6.1 Focus on specific EDM objectives

can be utilized (and briefly considering the related technical aspects), we can now consider the corresponding costs and financial issues.

These will be considered in three separate sections:

1 Costing the benefits arising from the successful use of EDM.
2 Outlining the direct and indirect costs of purchasing and operating an EDM system.
3 Financially justifying investment in EDM technology.

Costing the benefits arising from the successful use of EDM
Certain examples of the kind of savings arising from the successful use of EDM are included in an overall manner in the worked example later in this chapter. However, it is worthwhile here to briefly revisit the intangible (or should it be indirect?) benefits arising from EDM. The importance of these can be demonstrated by considering the adoption of computer-aided design (CAD). Although companies have always traditionally justified CAD on the basis of drawing office savings, in the majority of engineering companies the cost of operating the drawing office is less than one per cent of the cost of sales. This means that if a company can use CAD to increase sales by only one per cent, they can make more profit from overhead recovery than any savings they could realistically make in the cost of running the drawing office.

Fortunately, the magnitude of these intangible/indirect benefits means that accurate estimates are not needed because, for the correct application, even conservative estimates can show an attractive investment return. Additionally, the advantage of using these benefits is not just that they help to make investing viable, but that managers throughout the company who have been involved in the selection process will ultimately have a good understanding of the way that the particular technology concerned will affect the total organization.

Outlining the direct and indirect costs of purchasing and operating an EDM system

Direct costs

- Hardware

Server processor	Workstations
Client processor(s)	PCs
Disks (magnetic and optical)	X-Terminals
Memory enhancements	Backup/archiving devices (e.g. Exabyte tape)

    Network costs (if not already established—e.g. cards, cabling, hubs, bridges, etc.)
    UPS—uninterruptible power supplies
- Software
    Operating system
    RDBMS
    EDM software (perhaps on a modular basis)
    Networking software (if not already established)
    Bespoke/custom software and associated tools
    Application software
- Services

Training	Consultancy and services
Project management	Commissioning
Auditing	Testing
Acceptance	System evaluation costs
Installation	

Indirect costs

The indirect (or ongoing/recurring) costs of running a comprehensive EDM installation are usually related to the services aspects, but could of course include new hardware or software elements purchased as an ongoing requirement. This is an aspect many people forget or choose to ignore—a 'We've bought it now we can forget about it' approach. Ongoing costs include:

    Maintenance—both software and hardware
    Any upgrades not included under maintenance
    Possible support charges (e.g. hot-line option)
    Ongoing consultancy services
    Project audits
    System management and internal staff time
    System refinement and specification

Financially justifying investment in EDM technology

We considered the actual benefits which can be achieved from the successful use of EDM—at the strategic level, the mid and lower levels, and the intangible benefits—in Chapter 3. While we can now appreciate that many of the benefits are difficult to cost, companies cannot be forced to make significant financial investment decisions merely as acts of faith, even although they may be described as 'strategic decisions'.

The inability of advocates of 'flavour of the month' technologies to show that these fashions can be economically justified means that many companies have been faced with this very situation, or ignoring other projects that may be vital for their long-term survival. Unfortunately, experience has shown that many of the claims made for past technology investments were not achieved in practice and, long after its 'flavour' and associated advocates have gone, companies have been left with something of an expensive white elephant. Many managers approach investment appraisal along the lines of, 'How large a number do we need to establish to get the project past the accountants?' and, in practice, projects justified by this technique rarely achieve the planned savings. There are two main results which can accrue from such a scenario—a loss of credibility for the manager concerned and a growing belief that the technology concerned is a failure.

Traditionally, investment in technologies such as EDM has been difficult to justify financially, and until recently it was believed that such investment could not be evaluated using conventional techniques. It must be remembered that any investment decision is based on assumptions regarding future events and as such is therefore subject to error—it is not an exact science. One of the main problems with investment appraisal is that most companies have not updated their procedures to reflect the way technologies such as EDM are able to impact manufacturing. This can be seen by the continuing use of the simple payback calculation,—i.e. 'How soon will we recover our original investment?' Certainly the use of the payback method completely fails to take into account many of the facts which need to be considered when evaluating any form of advanced manufacturing technology. Reliance on this technique will guarantee a steady decline in investment in many organizations by its concentration on pure 'investment recovery' only. Sadly it is still the main form of financial justification in over 50 per cent of UK companies today.

However, recent thinking in the field of investment appraisal suggests that computer systems and software such as EDM can be evaluated in a similar way to, say, machine tools, and that it is possible to compare investment in different areas using the same criteria. The main problem with payback is that it takes no account of the timing of the cash flows or what happens after the end of the payback period. By projecting cash flows, a more realistic view can be taken to overcome past problems resulting from the way engineers have traditionally applied such principles to complex projects, and their inability to relate the technology to terms which accountants can understand.

Investment appraisal is normally only used to ensure that an investment will ultimately prove to be profitable. However, in practice it can be shown to be a technique which has a number of objectives, such as:

- To assist in identifying the most profitable areas of technology
- To ensure that investment in any chosen area will prove profitable
- To help identify the objectives and timescales of the implementation
- To help choose between different specifications and manufacturers/vendors
- To quantify the costs and benefits to ensure the investment is correctly reflected in the company's financial/costing system

The problem of making rational decisions about investing in EDM is made worse by the lack of consensus about its definition. This has resulted in many companies considering its introduction using purely subjective criteria when deciding what the objectives are, how to implement it and how much to spend. Clearly a balance is required between quantifiable and subjective criteria. The consequences of considering only subjective elements include:

1 Insufficient commitment of resources, or companies may concentrate on trying to achieve minor benefits at the expense of the major ones originally identified.
2 The progress of the project cannot be measured in terms understandable to senior management if quantifiable objectives are not included. This can ultimately lead to a withdrawal of management support.
3 If companies are unable to adequately define the benefits they are aiming to achieve, then they will also be unable to identify the most appropriate techniques to use.
4 Like many other techniques, EDM can be introduced as the 'fashionable' thing to do. As a result there is a danger that it could be abandoned when partially complete or 'unfashionable', as has happened with other techniques in the past.
5 Some organizations that could achieve considerable benefits from EDM may fail to introduce it because of apparently high costs and the inability to show it will be profitable. Conversely, other organizations may concentrate on EDM at the expense of other aspects of their operation that should have a higher priority.

Fortunately, we can now see that almost every benefit which can be identified can be redefined and quantified, and included in an investment appraisal. It can then be considered more as an 'indirect' rather than 'intangible' benefit. Indeed, Dr Peter Primrose of the Total Technology department at UMIST goes further. He quotes, 'No benefit should ever be excluded on the grounds that it is intangible' and he promotes the concept that all benefits are redefined into quantifiable terms for detailed analysis.

Payback calculations can be more easily applied when considering simple document image processing (DIP) or Engineering Document Management systems where the solution is more focused and not so all-encompassing. Take for example the simple situation of an operator whose task is to retrieve and print various documents in an engineering design or DO environment. Assume that he or she can process 25 documents per day and is paid £200 per week. In broad terms then each document costs £1.60 to process. If a system was purchased for this operator alone which increased efficiency by 50 per cent (a conservative estimate) at a cost of £5000, then each document would cost £1.07 to process. The payback period for the machine would therefore be less than one year assuming a stable workload; that is:

$$\frac{5000}{1.60 - 1.07} = 9434 \text{ documents}/187.5 \text{ documents/week} = 50.32 \text{ weeks}$$

Short payback times such as this are not uncommon when considering the increased efficiency which can be achieved, especially when labour costs account for a reasonable percentage of the overall figure. This can provide a sound basis for any justification. However, as the scope of the system is increased up to a fully functional EDM system such simple criteria cannot be applied and techniques such as average return on investment (ROI) and discounted cash flow (DCF) require to be used. DCF can use either of two techniques: the net present value (NPV) or the internal rate of return (IRR). Although the NPV approach is simple and convenient when ranking alternatives, the IRR is more easily interpreted by managers and is, in general, more widely used. In both cases the real problem is establishing the rate of return appropriate to the project concerned, since both of these methods can be considered tactical (i.e. shorter term) or project related, and not strategically oriented.

The implication in much of what has been said so far in comparing payback and DCF is that payback tends to concentrate on short-term projects and DCF on medium to long-term projects. Advocates of payback argue that the uncertainty of the economic climate inclines companies to adopt a conservative short-term approach, thus encouraging them to use payback. In practice the difference is shown to be more fundamental—the use of payback usually results in the wrong decision being taken.

Figure 6.2 shows an often implicit assumption that is made about payback-based savings—that a project will start to generate savings as soon as it is commissioned. Unfortunately, although investments such as EDM may be operational in a relatively short time, it is likely to be a long time (possibly

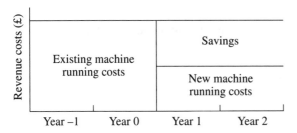

Fig. 6.2   Implicit payback-based savings assumption (*courtesy of Dr P. L. Primrose*)

many months) before they are producing maximum savings, although companies that use payback will normally reject any project that does not give a one (or possibly two) year payback return. Figure 6.3 shows a simple view of the build-up of savings over a two-year period. Although savings have reached the 50 per cent level by the end of the first year, the average for the year is only 25 per cent and the average for the second year is 75 per cent. If we take an example of an EDM system costing £100 000, a company looking for a two-year payback would only accept the project if savings were going to be £50 000/year and the start-up situation was as shown in Fig. 6.2. If, in practice, the savings accumulation was as shown in Fig. 6.3, the project would need savings of £25 000 in year one and £75 000 in year two to provide a two-year payback—which implies that the rate of savings in year three, and onwards, will be £100 000/year.

Using DCF (and assuming a ten-year working life) and a start-up situation as shown in Fig. 6.3, the £100 000 investment would only need to generate annual savings of £25 000 from year three to provide an annual rate of return of 16 per cent. Annual savings of £20 750 would provide a similar return if the start-up was as shown in Fig. 6.2. Although this would be acceptable to many companies that use DCF, it would be completely unacceptable to those that use payback.

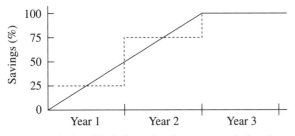

Fig. 6.3   A simplified view of savings accumulation (*courtesy of Dr P. L. Primrose*)

Although each case is specific and has to be justified on its own merits, the following example serves to outline a simple cost justification for the company as profiled below which expects five simple financial benefits to accrue from EDM. The actual percentage improvement figures would also need to be estimated via a suitable analysis.

COMPANY PROFILE

Annual sales	£20m
Cost of sales	£10m
Purchases	£5m
Indirect/clerical labour cost	£1m
Inventory	£5m

FINANCIAL BENEFITS

1 **Inventory** Assume 10 per cent reduction in stock (using 10 per cent cost of inventory):

$$£5m \times 10\% \times 10\% = £50\,000$$

Reasons – Less obsolete inventory as a consequence of improved engineering change control
– Faster reaction to needs.
– Less duplicate stock

2 **Customer service** Assume 5 per cent increase in sales at 8 per cent net profit:

$$£20m \times 5\% \times 8\% = £80\,000$$

Reasons – Increased ability to react to changes
– Better after sales and service aspects
– More effective sales teams
– Improved products/reduced costs

3 **Indirect labour productivity** Say 10 per cent improvement consolidated over all departments:

$$£1m \times 10\% = £100\,000$$

Reasons – Less unproductive time spent searching for data
– Engineers spend more time on what they were employed to do
– Improved communication
– Improved utilization of existing systems via integration

4 **Reduced physical media** £10000
Reasons – Increased terminal access
– Reduced errors
– Reduced paper reproduction and storage costs

5 **Purchasing**   2 per cent improvement in purchase costs

$£5m \times 2\% = £100\,000$

Reasons – Less re-engineering of products due to better control
– Improved traceability
– Better procurement policies and increased confidence in data from the system

TOTAL TANGIBLE BENEFITS PER YEAR = £340 000

COSTS

	Initial purchase	*Ongoing*
Computer software/hardware	£100 000	£20 000
Education	£30 000	£8 000
Implementation	£20 000	—
Consultancy	£20 000	£2 000
TOTAL COSTS	£170 000	£30 000

Return on investment

$$\text{ROI} = \frac{\text{Annual benefits} - \text{ongoing expenses}}{\text{Initial costs of implementation}} = \frac{340 - 30}{170} = 182\%$$

Cash flow

Year 1	−170 000	Initial cost
Year 2	+340 000	Annual benefits
	−30 000	Ongoing costs
	+310 000	Year 2 total
Year 3	+340 000	Annual benefits
	−30 000	Ongoing costs
	+310 000	Year 3 total
Year 4	+310 000	Year 4 total
Year 5	+310 000	Year 5 total

Five year cumulative cash flow   +1 070 000

Cost of delay

$$\text{Cost of delay} = \frac{\text{Annual benefits} - \text{ongoing expenses}}{12 \text{ months}} = \frac{340 - 30}{12} = £25\,800/\text{month}$$

These simple results are interesting and perhaps could act as a basis for a brief overview of what EDM could do for your own company. The ROI of 182 per cent is particularly interesting—if a machine tool had such a figure you would buy it immediately! Yet most people have difficulty doing the same for a technique such as EDM, even though the bottom line results are the same.

6.5   Decision time

At some point in time the critical decision requires to be made: are we going to adopt EDM or not? Will the advantages to our organization justify the cost in implementing EDM or are there other requirements that should take precedence?

This is a difficult decision to make and, like many projects of a similar nature, may well be hedged a number of times, either directly by the board or indirectly by other stalling techniques such as raising further questions or requesting additional reports.

Such actions cannot, however, continue indefinitely and a go/no-go decision will eventually have to be taken at board level. This review meeting can be considered as a 'commitment' review where each element of the project should be summarized and considered in turn as part of the process of authorizing the project. These elements can include the risks, costs, benefits, resources, etc. which will be required, and are very often included as part of a structured methodology. An example of considering an EDM project's review is shown in Fig. 6.4. This is shown as a single process but can be linked to others and a complete 'roadmap' of the overall project obtained.

Assuming a positive decision is made, how then do we proceed? What are the selection and implementation issues that must be considered? These will be fully addressed in the next chapter, and they rely directly on the outcome of the board's decision. For example, many companies choose to adopt a 'toe in the water' approach, where a decision is made to adopt a certain aspect or subset of EDM, for example file management, to prove the technology before proceeding onto the next. While it is appreciated that any technology which is new

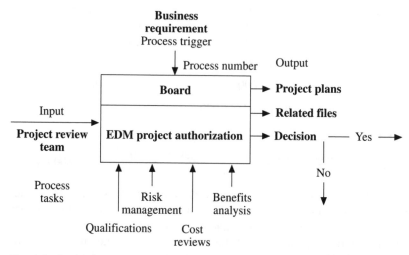

Fig. 6.4   Project/process control—a board decision model

to the people concerned will be treated with caution, in many such cases the half-measure system is subsequently accepted as being what was originally required and, as enthusiasm and commitment is lost, other more pressing requirements are identified and the full EDM solution potential is never actually achieved.

It is therefore important that the decision made by the board relates to the adoption of all relevant parts of EDM to the organization and that those who are responsible for making the decision understand the benefits and the nature of the task that lies ahead. At this point it should be noted that, although the investigation thus far will have involved obtaining information on EDM and contacting EDM suppliers, the board decision at this time is to approve the acceptance of going for EDM in general and to obtain the 'buy-in' to the concept at the highest level. An outline indication of requirements and budgetary figures will generally have been gained and a small project team set up, but the generation of a full requirements specification and detailed quotes from vendors only needs to be undertaken after approval is gained. After all, the decision may be negative. A full guide to the procurement and implementation issues which follow from this stage is shown in Chapter 7.

It is vitally important that the appropriate levels of management leadership and commitment are shown, both initially and throughout the duration of the project. Such commitment, from the highest level, will have an important effect on the ultimate success of any IT project and has been proven many times in the past. If the business does require to be re-engineered as part of an EDM implementation, then attitudes and culture will also require to be re-engineered, and that must start from the top. The board must ultimately own the whole process and drive it through: the EDM ideal of one system for data requires one vision to drive it through.

In many cases, although the decision to implement all aspects of EDM has been made, a structured and sequential implementation plan may be adopted with each particular aspect of EDM (e.g. file management, change control, configuration management, etc.) considered on a trial (or 'pilot') operation initially before acceptance and live use is made of it. This is an ideal way of subdividing and managing what can be a complex and time-consuming project in many larger companies, while at the same time working in the full knowledge that the entire concept has been agreed and that each element will eventually all fit together into a cohesive and comprehensive solution.

Like many similar technologies, EDM is not a panacea, and a correctly managed EDM justification is required and should provide senior management with the necessary information to make an informed and confident decision regarding investing in such technology within their own environment. In this way it is hoped that the correct decisions will be made by those companies which can successfully utilize the technology, as well as those who ultimately feel it to be inappropriate.

7

Implementation aspects

7.1 Where to start?

Because EDM, in its widest sense, operates over multiple functions, departments and divisions within a business enterprise, it is very difficult to visualize how or where to start while maintaining an overall view of where you want to get to eventually. However, lessons can be learned from other 'cornerstone' technologies of computer-integrated manufacturing.

The first task which needs to be undertaken is to develop a strategic vision— to determine the specific benefits which you seek to match your competitive need. This is often handled via a structured workshop for senior management and would normally be performed prior to identifying EDM as a specific and identifiable element of this vision. With regard to EDM in particular a thorough investigation of the current situation is initially required. You cannot achieve goals which have not been clearly set and you cannot set clear goals until you know clearly what the problem is you are trying to solve. This review of present procedures and systems must be conducted before the purchase or generation of an EDM application and should incorporate (or at least have considered) the aspects of procedural and/or cultural changes which could (and very often should) be made to the existing internal systems.

A company may be used to having its financial functions audited but when it is forced to scrutinize its whole approach to design and development, process management and internal communications, it often finds that the changes in general working practice, rather than simply adoption of the raw technology, also have a great effect on development cycles and time to market. These changes could be introduced either before, or as part of, the EDM system introduction itself. Indeed, I have seen many companies abandon their plans to purchase a system after this initial stage and still reap significant rewards from studying and improving existing procedures. These aspects of bending a system to suit the requirements and vice versa was discussed in Section 4.4 and is an important element—very often considerable cost savings can be made by the adoption of simple alternative re-designed processes.

Items to be addressed during this review are of course company specific, but can for example include:

- Money tied up in underutilized data storage facilities
- Volumes of re-keyed data
- Ratios of time spent searching and manipulating data by various departments (especially design/drawing office)
- Assessment of existing company procedures

Not only does addressing these kinds of issues focus on the associated problems, it should also provide data which can be used for reference purposes after a new system is introduced. Items such as the company's installed computer systems, data volumes, ownership and usage, flow procedures, security, expected growth, communication paths, etc. should all be studied. External consultancy services may also prove to be extremely useful at this stage. Many companies have also found the use of an EDM 'health check' matrix specifically aimed at ECOs (Engineering Change Orders) has also proved to be of great benefit in highlighting those areas where improvements are required. An example is shown in Fig. 7.1.

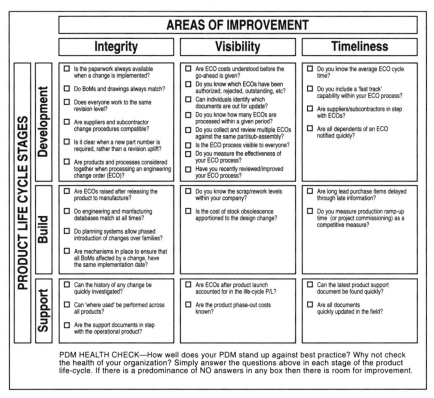

Fig. 7.1   EDM 'health check' matrix (*reproduced with the kind permission of Keith Nichols*)

Some of this information may already have been generated from the work done to acquire a better understanding of EDM and gain board approval for the go-ahead to source and implement the technology. If, however, this initial work was undertaken some time ago it should be revisited to confirm its accuracy—such technology moves fast and procedures may have changed, for example. Initial information and technology awareness can come from a variety of sources, for example other users (sharing experiences and information), vendors by honest (since a long-term relationship may result) exchange of information, trade shows, magazine articles, etc. One of the first and most important tasks is to set up an environment which will involve all relevant persons from the main board downwards in a continuous programme of awareness, communication and education—the 'people' side of EDM. This is of fundamental importance. Education and involvement are required but both terms can be misunderstood. A person can be involved in implementing EDM without accepting responsibility for the success of the venture. Caution is required when using the word 'involvement'—the 'system users' must accept the responsibility for success, since only they can make it produce results. This means that the users (and this implies a range of users from the Engineering Director to the Drawing Office clerk) must understand the technology in question, why it is required and how it operates. This requires education—relevant education at all levels. Education must precede intelligent participation: training comes later and is usually system related.

There are two main elements to be addressed here:

1   Fact transfer—to allow the what, why and how questions to be answered.
2   Behaviour change—educating people out of their existing informal environment into one where they become convinced of the need to do their jobs differently.

These two points are very important and contribute greatly to the success or otherwise of many installations: education is a major part of implementing EDM and should be carefully planned. Performing education effectively is synonymous with managing the process of change. It is important that personnel receive education that is adequate and appropriate to their needs and, at the same time, consideration should be given to the cost of this education.

Also, following on from the initial phase of obtaining board approval, a project team needs to be set up to formally generate a requirements specification or invitation to tender (ITT) and to evaluate and shortlist those potential suppliers for eventual main board approval.

At this stage the project leader and project team do not need to be as formally structured as required during the implementation stage (assuming go-ahead is given), but clearly defined roles and responsibilities do need to be established and procedures defined. The choice of project leader usually causes some concern at this stage—should it be an IT or engineering function?

It is important to agree an early consensus between these two areas at this stage: both are going to play key roles in the implementation so differences need to be settled early. Most experienced users agree that engineering should take the lead at this stage since it has a daily responsibility for the manufacturing issues which EDM addresses. In addition, all members of the team need to remember that, without a common purpose, the project will suffer.

It should also be borne in mind that in the interests of continuity and relevance of expertise, some members of the evaluation team will continue through to the implementation team and may be off-line from their 'normal' work for some considerable time. Indeed, I have seen many cases where careers have changed as a result of this kind of exercise. Such a volume of knowledge, background information and enthusiasm is generated by certain team members that returning to their prior role within the organization would be counter-productive. Appropriate plans therefore require to be established to cater for such an eventuality and a recognition made that the success of the eventual solution requires the provision of suitable resources which may involve the temporary or permanent re-assignment of staff.

In summary then, starting out in EDM, like many other aspects of a corporate nature, is not easy. The first stages of an implementation are often the hardest. They involve steps into the unknown, yet often have the highest profile inside the organization. There is no 'correct' way of doing it. However, discussions with those companies which have successfully initiated EDM solutions shows that there are six main rules to be followed. They are not exhaustive—the whole issue is too complex for that—but there is a wide consensus that they are essential stages on the road to successful implementation. They are:

1 Examine your motivation in looking at EDM and prioritize the requirements you have previously specified. This aspect is not related to individual systems functionality—these are business-related issues here. Perhaps the use of a suitable business modelling tool may help.
2 Understand the nature of your organization and the implications this imposes on your choice of solution—the use of standards and how the proposed solution matches this need, for example. It is also important to consider the size and experience of the supplier: will it be capable of supporting a large corporate user? Can it customize the solution if required?
3 Evaluate possible solutions carefully. Look for the base functionality you have previously identified, not extra frills you don't need.
4 Build and test a pilot system having first defined the scope of the pilot implementation to ensure the original objectives are met. Pilots are discussed in more detail later.
5 Get the fundamentals correct at the pilot stage—don't leave them till later. Modify and enhance the pilot system until live operation is achieved.

6 Do not ever assume you are finished. A successful pilot may well lead to a company-wide implementation. Many EDM projects have proven that the process is addictive. Some companies have started purely with a wish to handle engineering changes in a more structured manner, but have continued the process by implementing configuration control, interfacing to MRPII and the rest of the downstream processes. Having built the critical foundations successfully, the prospect of managing the entire life-cycle of a product seems less daunting, and the benefits more clearly apparent.

7.2 Sourcing a system

The information generated from the initial reviews of EDM can be supplemented by further work by the evaluation team. This data can then be used in the generation of a requirements specification—a list of functionality required to overcome the deficiencies of the present system and to cater for the future data management needs of the company. A typical requirements specification usually lists the software functionality required from the proposed system. This specification of requirements can then be utilized in the system selection process.

Most companies involved in the selection of a major computer-based system perform their selection process in three distinct phases:

1 *Initial phase.* Vendor/system research, via trade magazines/exhibitions/market research, etc. in order to produce a 'mailshot' to a wide range of possible suppliers. This phase need not include the provision of a full requirements specification—a three/four page overview of requirements is all that is needed at this stage to allow the vendor scope to obtain an overall feel of how his system can best suit your requirements.

2 The production of a *shorter list* from stage 1 of systems which are felt could do the job—perhaps 8–10 suppliers. At this stage, an initial visit could be made either by the company or vendor to assess requirements prior to consideration of a more detailed requirements specification. This stage is usually accompanied by a demonstration/visit to view each of the suppliers/ systems concerned for initial proof that the system will perform the necessary tasks.

3 The preparation of a *true short-list* of up to three (maximum) final contenders. A further demonstration will be required and additional proof given to confirm the suitability of the system. This stage should result in the proposed system being presented to the board for approval. The project team should recommend the chosen system via the executive steering committee.

These points are noted in more detail in Fig. 7.2.

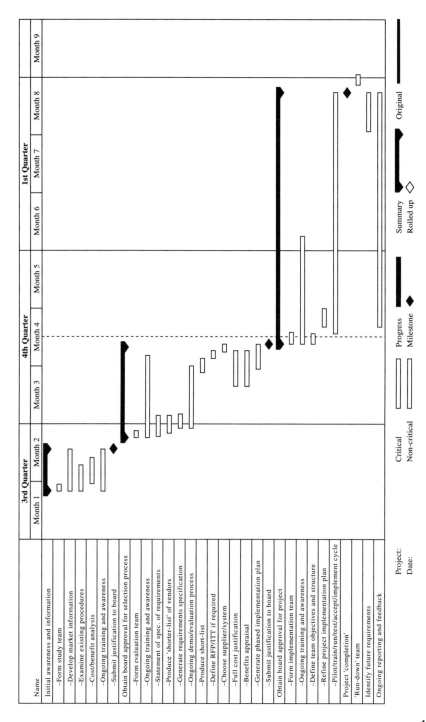

Fig. 7.2 Outline EDM selection process

145

Specific aspects relating to the selection of an EDM system include:

Initial contact

This should be kept to a realistic level. Systems and suppliers who you feel unsuitable should not be included just for the sake of it. Many of those contacted may also wish to visit at a very early stage—this can be worthwhile but also very time consuming if encouraged.

Where to look

Many people who are new to EDM find this difficult initially, even when searching for basic information for board consideration. However, this should not prove too great a problem in the long term since, once the word gets around that you are in the market for a system, suppliers will beat a rapid path to your door and you may find yourself swamped by all types of vendors claiming to provide exactly what you require. The EDM market, because of its evolving and somewhat fragmented nature, is a complex one. One major problem is that most EDM vendors have, until recently, addressed the market as an adjunct business opportunity and have not invested sufficient resource to establish a strong leadership position outside their traditional installed bases. Here are a few profiles of the major suppliers of EDM products:

1 *CAD vendors*   These suppliers are the earliest entrants on the EDM scene and supply a wide range of systems. Because of this they have built up a comprehensive knowledge base (although it may be biased towards the design/DO environment) and will generally offer consultancy services to integrate and implement their solutions.

2 *Independent EDM suppliers*   These suppliers are growing in number as the market develops and are able to provide solutions in an 'independent' manner, i.e. equally able to cater for design, manufacturing, file management, document image processing, interfaces, etc.

3 *Major platform suppliers*   The changing nature of the computer marketplace has meant many of the large system suppliers taking more interest in 'solutions' such as EDM than their previous hardware bias would allow. Companies such as IBM, DEC, ICL and Bull have all invested in EDM solution technology and have a wide range of solutions between them.

4 *Image and document management suppliers*   Such systems have been on the market for some time now (e.g. Xerox Document Management Systems) and initially some confusion may have existed between them and EDM. Such technology (perhaps more commonly known as DIP or Document Control Systems), as we have seen, now forms an integral part of many EDM solutions. These suppliers are increasing the functionality of their offerings and are becoming a more significant force in the EDM scene, although they may exhibit a lack of engineering focus.

5 *Manufacturing/MRPII suppliers* Because MRPII systems are now expanding in functionality and embracing the COMMS concept (Customer Oriented Manufacturing Management Systems), certain suppliers are now expanding their offering into the area of engineering change control and configuration management. This is an embryonic but growing area, especially in Europe.

6 *Systems integrators* The providers of SI services are growing and are able to provide the skills required in dealing with projects of the nature of EDM. For example, legacy systems require to be interfaced and integrated, program management skills apply, suitable team resources are required and enterprise-wide considerations apply.

The cost of EDM products from whatever supplier source varies immensely, just like the associated functionality. The current range is from around £10 000 to over £100 000. Much of the divergence depends on the extent to which EDM is used, the functionality supported, and the level of potential benefits likely to be delivered.

Requirements specification

As we discussed briefly in Chapter 6, a requirements list or specification is needed to help focus the often diverse ideas on needs and requirements from each company department. This is simply a list of functionality required from the EDM system, based on the results of the previous initial review stage, and identifies a company's information needs and required system functionality. The temptation often exists at this stage (in a similar manner to when perhaps the existing procedures were being documented) to design the system. This is not the objective here and the list should be kept reasonably concise and factually correct. A determined effort should be made to keep the requirements as standard as possible—if for example, a rather convoluted method has developed over the years and no-one can reasonably prove why it should continue, then it should not be specified as being required.

Without a requirements specification, any evaluation team gathered to consider the use of a computer system will quickly degenerate into a group of individuals eager to pursue their own interpretations of what is actually required. I well remember one such example in the early 1980s, where a team of engineers were reviewing an advanced (for that time) 3D solid modelling system. One of the senior engineers in the party, well known for his individuality, became increasingly upset during the demonstration that the system did not keep up with his impression of advanced technology. This continued until he could take no more and he exploded with rage when told that the system did not have the capability to attach sensors to his head and model any object that he cared to imagine! The other members of his party were required to calm him down and explain that this function, apart from not being possible, was

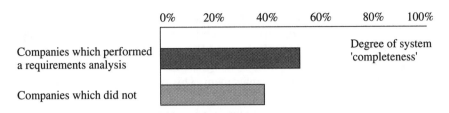

Fig. 7.3 The effect of a requirements specification on the completed system

not even envisaged as being required for their particular application. The sad part about this story is that the person in question really believed that this functionality should have been both achievable and available at the time, and although it is perhaps an extreme example, serves to show what can happen when people's views are not aligned to a common set of goals and requirements.

In summary then, the writing of a requirements specification is seen as a very important aspect in introducing an engineering computer system such as EDM into an organization. Failure to effectively conduct this analysis often leads to implementing a less complete system solution by concentrating on non-critical aspects, as is shown in Fig. 7.3.

There is no 'standard' layout of requirements specification: it is very much a company-specific document. The following, however, provides a framework of a 'typical' specification:

1 Your company-overview, markets, current situation, outline of future plans, etc.
2 The products produced—considerations thereof
3 Project review, objectives and requirements
4 Details of any existing computer systems
5 Pertinent data and volumes
6 Required functionality (hardware and software)
7 Integration aspects (with existing systems)
8 Request for vendor information (locations, stability, training, users, consultancy services, experience, indication of likely cost, etc.)
9 Possible future functional requirements.
10 Due date for the proposal and relevant contact person

Type of solution

Following on from Chapter 6, we require to address the issue of purchasing a standard software package or having one specifically written. Choosing the correct system will involve the requirements specification, ITT/RFP and a number of other decisions. The preferred system may have to cope with many changes which are not yet clearly defined. For example, perhaps our present marketplace requires us to employ a design-oriented approach where a DOMS or Engineering Document Management centred system would suffice. The future direction of the company (or general market trends) may result in this changing to a more product-centred environment where the emphasis would be placed on configuration management. The chosen solution would require to grow in line with such changes and it is important that it can do this both horizontally and vertically throughout the organization in as seamless a manner as possible. A decision will also be required on the amount of software customization which is to be included in the final solution. This is a different question to the amount of 'tweaking' of the solution, perhaps using the tools and facilities provided with the chosen solution, to suit the required application. Most systems, in recognition of the fact that very few manufacturing companies operate in exactly the same way, now provide such facilities to enable these changes in operating procedures and standards to be accommodated without changing the actual program code of the EDM system itself.

This decision focuses on whether a standard 'shrink-wrapped' software package would suit the required application, or whether a totally custom (bespoke) solution is required due to the unique nature of the requirement. These two approaches are diametrically opposed, as shown in Fig. 7.4. Like many computer-related projects, most early (and some current) EDM systems were heavily tailored or completely customized solutions written around an organization's existing procedures and operations. For example, the US helicopter manufacturer, Sikorsky, has so far spent $77m over the past seven years developing its own EDM system.

One alternative here could be to employ a software house or EDM supplier to change the system code, thus avoiding the direct cost of in-house software

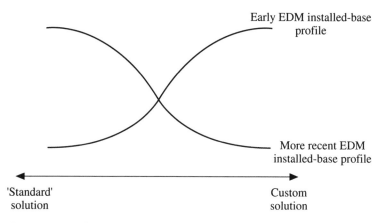

Early EDM installed-base profile

More recent EDM installed-base profile

'Standard' solution

Custom solution

Fig. 7.4 EDM installed-base profiles

development. However, functional flexibility can be lost in the future, and the customer remains dependent on the supplier for further future changes. Over the past few years a recognized market has been identified and some common tools and utilities established which EDM systems can utilize relatively easily. More 'off the shelf' EDM solutions have been introduced (although early systems of this type did not provide the tools necessary to successfully tailor or 'tweak' the system to suit individual needs). Because this trend is now reversed, a more preferred situation exists where all systems have reasonably comprehensive toolsets, and the bias of the installed EDM user-base has moved accordingly, as can be seen.

In many respects EDM systems have evolved in a similar manner to other IT solutions where the initial trend for custom solutions has given way to standard applications. Although specialized computer programs provide exactly what is required, they generally prove more expensive in the long term. The currently accepted general trend is to keep a standard core of system functionality for ease of use, upgradeability, etc. and to customize only those 'frills' required.

Requests for proposal (RFP), invitations to tender (ITT) and requirements specifications

We can see the relationship between these in Fig. 7.2. In this case the ITT/RFP and requirements specification are taken to mean different things, but many people consider them to be the same—there are no hard and fast rules. The RFP and ITT are generally considered to be different terms for the same document. In Fig. 7.2, the ITT is considered to be a more detailed version of the requirements specification and would generally include more information on exactly what was required and why. It would certainly request a detailed cost breakdown of the proposed solution. It may also include some form of identification to indicate the relative importance of each topic—for example three groupings such as mandatory, desirable or useful. A sub-option to these is to consider grouping the second two with a timing element. For example, if function XYZ is not currently available, then a projected availability of 1st quarter 1996 is acceptable. If such aspects are important they should be included as part of the contract. An effort should be made to avoid the common tendency to list as mandatory every conceivable feature—this drives up the price considerably and many of the features may never be fully utilized.

Whether answers to questions have been mailed or discussed face-to-face, the problem of assimilating and ranking the results remains. How should this be done? Techniques vary from an in-depth ranked score to pure 'gut feel', the ideal being somewhere in between. It is very often easy to get carried away with the exercise of choosing the system rather than getting it installed and making it work. It is certainly useful to analyse the results in some form—hence the reason for defining the mandatory, desirable and useful options. But there are also a lot of subjective criteria to be borne in mind (which should also be noted in writing): ease of use, look and feel, robustness, documentation, people concerned, vendor company culture, etc.

Obtaining proof

In any computer system sale there is a large element of proof involved—proof that the system will suit the previously defined requirements. These aspects are briefly outlined below and covered in more detail in Chapter 8.

No-one will purchase a system without adequate proof that it will indeed perform the required operations. However, it is very easy to overplay this aspect and attempt to investigate every small element of a system, even to the detail of including certain custom development work. This can result in a long evaluation period—an undesirable situation for all concerned. Adequate proof can take many forms, and each situation will merit its own selection of those which are relevant. They include:

- An overview demonstration
- A telephone call to an existing user
- Trial use of the system
- A visit to an existing installation
- The information from technical articles
- Product brochures and documentation
- A system benchmark exercise
- A response to an ITT

7.3   Setting objectives and project authorization

As with any major undertaking, clear terms of reference for the project and those involved need to be established. These should be documented and agreed alongside the objectives of the project. If the objectives of introducing EDM are never documented and agreed, then how will we ever establish when the project is complete and how successful it has been?

Each organization's list of objectives will of course be unique but will in general not be too difficult to produce. This is due to the fact that the problem areas will be well understood already (they were probably one of the main driving forces when considering EDM initially) and the exercise of measuring the data volumes as part of the requirements specification/ITT generation will further highlight these areas requiring attention. Objectives could for example include:

- Reducing inventory levels by 10 per cent within an 18 month period
- Streamlining the change control process from X weeks to Y days

- Reducing average waiting time for drawing/information retrieval from A to B
- Improving service levels by 20 per cent
- Cutting the number of monthly concessions to Y
- Improving purchasing performance/vendor scheduling by X per cent
- Obtaining a correct PL/BOM 'hit rate' of 99 per cent

This overall list may be mapped onto a refined list detailing primary and secondary objectives. The primary objectives should be qualified and clearly defined for all to agree. They should also have timescales set and incorporated into the master implementation plan as milestone events. In setting these objectives though (and they should be included in overview form as part of the specification of requirements/ITT to enable vendors to clearly understand your requirements), another question arises. How should they be prioritized? Which objectives should be aimed for first: the largest or smallest?

Here again opinions vary, but once again lessons can be learned from other projects of a similar nature. Clearly a balance needs to be maintained between setting the most important (and probably largest) objective first with its correspondingly long timescales, and the need to achieve some lower-level milestone event early in the implementation to maintain enthusiasm and confidence in the overall solution and the project's ability to meet all its stated objectives.

This situation serves once again to highlight the need for ongoing knowledge, awareness and education in EDM from the highest to the lowest levels in the organization. When this knowledge base exists, it is much easier to prioritize objectives and agree them with all concerned, rather than to spend a lot of time explaining to senior management why 'maintain configuration management records', although perhaps the most pressing need, should be last on the list and that a correctly structured file system and change control process should be established first as necessary foundations to build on. In a similar manner education and awareness should be able to rationalize an individual's lower-level objectives such as operators understanding why a database procedure definition and pilot is required to be established before they can obtain their most pressing requirement such as an obsolete stock listing or upcoming change note report.

Everyone involved should therefore be able to buy-in to the list of objectives and to appreciate how they should eventually all fit together into a cohesive and complete solution.

Many evaluation teams only concentrate on the chosen solution and supplier when submitting their justification for approval. These two points are of course vital and an ongoing dialogue between the board or steering committee (if you have decided to form one at this stage) and the evaluation team should ensure there are no last-minute surprises. It has been known for many

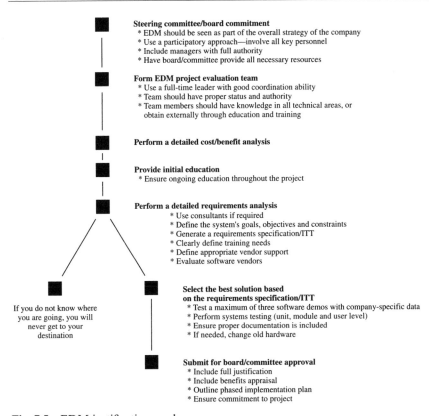

Fig. 7.5   EDM justification roadmap

months of independent work in evaluating systems to be wasted because of such a lack of communication: certain companies may have to choose from one supplier or a specified range of suppliers as dictated by the holding/parent/partner company—a fact sometimes not clearly understood or communicated to the evaluation team.

However, the other elements we have discussed to date are also very important and should not be overlooked. An outline review of these is given in Fig. 7.5.

7.4   Structures, reporting and responsibility

Let us move on now from considering pre-board/committee approval to concentrate on the implementation aspects involved in a post-approval situation.

However well planned, once the button is pushed and the decision made to introduce a specific EDM system, there will probably be a number of worried people, at least initially. What happens if the hardware doesn't connect into our existing network properly? What happens if the software modules are slow? What happens if it all doesn't work?

The key here, as always, is planning. If these eventualities are planned for and managed correctly (including the use of correct risk management techniques) success should be achieved. However important planning is, it must also be mentioned at the outset that the implementation of a complete EDM system involves a lot of work. To quote from Ed Miller, President of CIMdata, 'If EDM was easy, we'd have done it years ago.' But let us not be too pessimistic here—the fact is that the benefits are there and have been identified as being achievable, and we have a plan of how they will be obtained. A realistic approach is required to see it through.

The best overall framework for projects of this nature is a three-tier structure which in many respects can be considered as a more formalized version of the evaluation team involved previously to source the system (Fig. 7.6). The elements of this structure will now be considered.

The project leader

The project leader is the key person who should be in charge of the EDM project team and spearhead the implementation at the operational level. Choosing the appropriate person is therefore of crucial importance. Past experience has shown:

1   Wherever possible appoint a full-time project leader. If a part-time leader is appointed their commitment to the project will be reduced by having to spend time on normal duties. In smaller companies where the expense of a full-time leader cannot be justified, steps should be taken to ease the normal duties to allow at least the majority of time to be spent on the EDM implementation.

2   The project leader should be a company employee. Hiring an 'outside

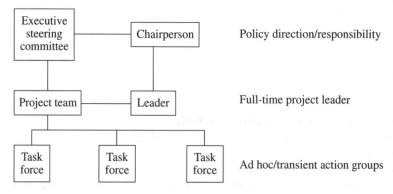

Fig. 7.6   Overall project structure

expert' to do the job has been shown to produce poor results. This is discussed further in Section 7.7.

3   The person chosen should have an 'operational' background, such as design, production, planning, etc.—such people tend to have more appreciation of 'real world' deliverables and the implications thereof. DP, accountants and sales, etc. tend to be more commercially biased and do not have this expertise. Selecting a DP person also reinforces the 'it's a computer system' message and reduces the ownership aspects of the project for the engineers and design staff involved.

4   Pick the best person available. This sounds obvious but try to choose the best 'all rounder'—someone with demonstrated abilities in leadership and communication, drive, knowledge of the company and EDM itself.

5   The project leader should be an established employee. If a new company employee is chosen then credibility and trust have not yet been established and will ultimately affect the outcome of the project.

6   Choose a leader with good communication skills. Since the project leader's job will be heavily involved with people, it is important to be able to explain, involve, enthuse, etc. all those involved.

The project team

The project leader heads up the project team, whose role is to implement all the various aspects of EDM at the operational level. Its jobs include:

- Establishing/maintaining a project schedule or plan for the installation and implementation of EDM
- Reporting actual performance against the plan
- Identifying problem areas and taking the necessary actions to minimize the effects of these (perhaps via sub-'task forces')
- Allocating resources to the project, assessing priorities, etc.
- Recommending and reporting to the executive steering committee
- Ensuring a smooth, rapid and successful implementation at operational level

The project team can consist of a few full-time members (should circumstances allow) and generally also involves part-time members from all relevant company departments. Project team meetings should be held regularly—perhaps once a week for one to two hours—and should be brief and to the point. Review status and actions, make decisions and allocate any new actions—no education or getting involved in too much detail.

The executive steering committee

This consists mainly of company top management and its function is to ensure a successful implementation. It should generally meet once or twice a month for about an hour to review the status of the project. Members should include general managers, directors, divisional managers/directors, etc. and the project team leader.

Early decisions include the overall scope of the project, approval of the project budget and schedule, and the staffing of the project team. Once the project is under way, the project leader should report progress relative to the schedule and be able to explain any shortfalls and actions/resources required to get back on target.

It has also been shown to be useful if one person on the steering committee can act as a 'focus'—rather like the project leader is to the project team. This person typically should chair the meetings and represent the committee/board. Key factors are enthusiasm for the project and a willingness to see it through—a top-level sounding board, flag waver, etc. Very often the project leader will report directly to the steering committee chairman—it is an important function but not necessarily time consuming.

Task forces

Certain actions that result from the project team are best handled by specially formed ad hoc groups who resolve the problem, feed back to the project team, and then devolve. This enables the project team to avoid getting involved in too much detail. Actions requiring a 'task force' could include structuring BOMs, loading data, rationalizing outstanding engineering changes, testing programs, etc.—in general, those tasks requiring 3 to 4 weeks effort at most.

Finally, the importance of holding regular meetings and documenting the progress of the project are elements which should not be underestimated. Many projects of a similar nature have floundered because some time into the project, when the pressure is on and it seems best to get the head down and get on with it, the meetings have been skipped or abandoned altogether. The resultant lack of communication, drive and planning is felt by everyone and as enthusiasm and commitment is lost, the project starts to spiral downwards. Those who are appointed to the roles previously outlined must carry the overall accountability for the results of the project. They must maintain project momentum by continuing to inform and involve all those concerned in an open and trusting manner.

7.5 Project implementation and timescales

One of the first questions usually asked by a company considering the introduction of EDM is: how long does it take?

If we could define when a project of this nature is complete then we could answer the question! What we can do is define when the major objectives have been achieved, though there will always remain those tasks, extensions, wishes, refinements, etc. of a legitimate nature, which can extend the time-scales considerably and in reality will probably always exist. It is important to set an achievable timescale target and stick to it, otherwise the project will continue *ad infinitum*. This should be around 9–16 months for the average system. If a 3–4 month timescale is set the chances of meeting it are reduced due to the volume of work which usually needs to be undertaken. In these cases a severe danger of running before the company concerned can walk is apparent. If objectives and timescales have been set and these are now known to be very ambitious (as a result of increased knowledge regarding EDM) the company concerned has to face the embarrassment of rescheduling them or of making them work (the course of action normally taken). The resulting short-cuts and pressures involved in such circumstances will mean a less than satisfactory solution is eventually obtained and the possibility of an extended project timescale anyway to address 'the last 10 per cent' of the project which always seems to be outstanding no matter what resources are committed to resolve them.

Similarly, if a project timescale of longer that two years is envisaged, the odds of a successful implementation are also shown to decrease sharply. It becomes more difficult to maintain the intensity of the project, the enthusiasm, drive and dedication necessary—all elements which should not be underestimated. The aspects of change must also be considered on implementation timescales—changes to the people involved (promotions, retirements, etc.) and environmental changes: for example a business slump makes the project too expensive and a business increase means people are too busy running the business to give adequate time to the project.

Even though a project timescale of X months is set for an implementation a schedule slippage will occur: many people underestimate the tasks involved. A small slippage is tolerable and should be planned for. A large one cannot be tolerated, because the benefits from EDM will take longer to achieve and will affect the cost justification initially put forward for the entire project.

It should also be remembered that there is no such thing as a 'standard' project timescale. Each one depends on the individual implementation concerned: the size of the EDM system being introduced, the applications that may already be in place, or the database structures already employed.

Whatever size or type of system is being considered it is generally best to subdivide the task into a number of major phases (usually three) which are tackled in a serial manner. There are also a number of individual tasks which can be undertaken simultaneously within each phase. This is shown in Fig. 7.7.

These phases and tasks can be structured to correlate to the objectives that have previously been defined. For example tasks 1, 2 and 3 of phase 1 may achieve the first objective of the EDM project and tasks A, B, C and D achieve the second objective. The system 'pilot' operations which will be discussed further in Chapter 8 also relate to this scenario. For example a pilot operation

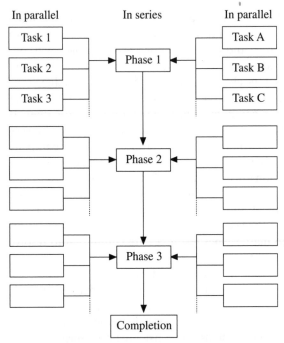

Fig. 7.7   Individual project phases and tasks

may require to be established to test tasks 1, 2 and 3 before live use. In this way pilot runs can be used to confirm tasks/phases which then relate to individual objectives.

Each of the main phases are very company specific but generally speaking could be split as follows:

Phase 1   Hardware integration, installation plan, database structuring, simple file management (vault) setup, initial project management.

Phase 2   Completion of file management system, workflow management including change control, integration of application modules.

Phase 3   Project management module, PL/BOM reconciliation, configuration management, full application integration.

When considering the actual software implementation it is important to differentiate between the appropriate business functions and software modules. Software modules are pieces of computer software which exist to support people in the effective execution of the relevant business function. Some implementation strategies are designed around the chosen systems software modules (i.e. a modular approach), for example project management, file structure, change management. This approach is popular with the solution vendors but does not have to be followed if it does not suit your environment.

It tends to focus on the system as a 'computer system' rather than an enabling/people system. We have discussed using 'pilot' implementation techniques thus far, so it is probably worthwhile outlining exactly what is meant here and considering the other traditional techniques of project implementation.

The parallel approach

This involves running the old system (if you have one) and the new system together for a while to gain confidence in the new system before unplugging the old one. The outputs are compared and when the new system appears to give consistently better answers, the old system is dropped. This is a traditional and popular approach—it proves to management that the investment is working and can be cost justified. However, it is expensive and difficult to maintain and operate two systems side-by-side, and is the old system's output accurate enough to benchmark against anyway? The parallel approach is fine if the current system works (e.g. accounting systems), otherwise it has been proved to be difficult to use.

The 'push button' approach

This is the exact opposite to the parallel approach and assumes that one Monday morning everyone starts using the new system in place of the old one (assuming one, in whatever form, existed), which has been removed over the weekend. A very high risk option. You can almost completely lose the ability to control your engineering data if you get it wrong—and there is no old system to plug back in after a few days/weeks of poor performance from the new system to retrieve the situation. Generally not recommended.

The 'pilot' approach

This technique involves selecting a group of products, or one, or indeed one part of a large product (as long as it contains a reasonable number of parts) and operating the new system on it, while controlling the other 99 per cent of the company's products in the normal manner. This proves that the system works before committing all operations to it. We will consider pilot system operation in more detail in Chapter 8.

7.6 Project management

Bringing in a project on time and on budget is still a nightmare to many experienced project managers. A recent UK survey by consultants KPMG Peat Marwick found that 62 per cent of all the companies surveyed had experienced projects running significantly over time, cost or both. In only 7 per cent of cases were technical issues the main cause of the problem—85 per cent of respondents mentioned inadequate project management, or management-related issues such as failure to define system objectives, lack of communication and unfamiliarity with the project scope, as the main culprits. In certain cases, changes or 'compromises' made during the pre-sales stages of a project may also ultimately affect its final outcome.

We can therefore see that any IT-related project such as the introduction of EDM requires to be managed correctly if it is to succeed. Although it is not the purpose of this book to detail the project management techniques and methodologies relating to such projects (the reader is directed to the Bibliography for further sources of information here), this section will outline some of the major elements which form the basis of successful project management.

The first point to remember is that managing a project is an art, not a science, and like many of the items we have discussed so far has no 'right' or 'wrong' course of action to achieve the eventual aim. However, there are a number of clear stages and procedures which can be adopted to form what is now recognized as a solid foundation on which to build any project. Many of these have been formalized into specific 'methodologies'. The project manager's role is to define, organize, plan and control the three main elements of any project. These are:

- The *scope* of the project. This defines the areas to be addressed in meeting the objectives, limits the project to these and defines the boundaries and relationships with other projects and systems.
- The *timescales*. To help with planning and control, projects are normally split into successive stages with associated timescales. It is important that these are kept realistic and achievable.
- The *resources* available. These are what the project manager has available in order to achieve what is required. They include, for example, the project team with their combined experience and skills as well as the physical resources such as project management tools, computer resources, office space and equipment.

These elements and the tasks which surround them are shown diagrammatically in Fig. 7.8, with the project manager often acting as the juggler to keep the three balls (and some say many more) in the air at all times.

A good project manager should be able to:

- Anticipate and avoid problems
- Tell people early what they need to do and when
- Plan thoroughly and balance workload, resources and timescales at all times

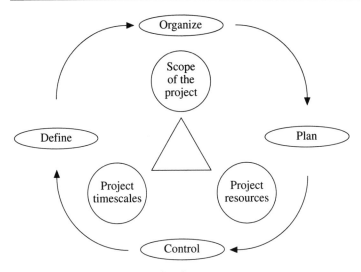

Fig. 7.8 Project elements and tasks

- Understand the scope and objectives of the project and always be conscious of these when making decisions
- Put the project first
- Meet and solve unexpected problems promptly and efficiently
- Protect the team from distractions and maintain the momentum of the project
- Relay any bad news to the steering committee promptly, along with practical options and firm recommendations

The task of defining, organizing, planning and controlling a project can be related directly to the activities outlined in Figs 7.8 and 7.9.

- *Defining a project* The business requirement is the reason for the project's existence. It is the problem or problems which need to be solved or the gap

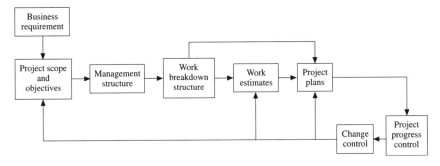

Fig. 7.9 Outline project stages

in the business which will be filled as a result of the successful completion of the project. The objectives of the project should clearly state how the business need will be solved and would normally involve any system design, targets for system availability, cost or performance issues.

- *Organizing a project*   Management responsibilities need to be apportioned in order to ensure that decisions are taken promptly and with the required degree of authority. Project organization also involves the generation of a work breakdown structure (WBS), which describes the project's elements and tasks, progressively subdividing them into more and more detail.

- *Planning a project*   This requires estimates to be made of the volume of work that each task generates for the project team and how much money will need to be spent on resources before they are completed. These estimates, together with the WBS, form the basis of the plans which determine when the project can be completed. They should show the sequence in which the tasks will be performed and their duration, and are matched to the resources available.

- *Controlling a project*   The project manager is required to know where the project stands at any given time; this applies to every individual element. Controlling a project also requires that changes are adequately catered for. Because individual elements of a project will change as the project develops and matures, it is vitally important that correct change control processes are employed to monitor and document these changes, just as the eventual software solution (i.e. EDM) will do for the engineering data! It is also important that the original risk portfolio, identified at the outset of the project, is managed and kept up-to-date.

Just like a product, each project has a unique life-cycle, and we have already seen how milestone events can (and should be) built into a project plan and related to the stages we have identified (refer to Fig. 7.7). Milestone events in a project plan, whether they are major or minor, are an important element in achieving eventual success. They relate to the achievement of specific objectives, and since they are visible and are clear targets to strive for, they can prove to be excellent morale boosters when they are achieved. For this reason one major milestone should be set fairly early on in the implementation phase. Their achievement should be communicated and celebrated by everyone concerned but the project momentum needs to be maintained—it is important to move on and aim for the others.

Project life-cycles will also contain decision points, again of a major or minor nature. These points can act as checkpoints or reviews within the project to confirm the project status and monitor the financial viability. The project risks also require to be reviewed at these decision points—perhaps changes which have occurred between them necessitate a major restructuring of the project, for example.

Finally, we all hear about the projects that go wrong—probably mainly by word of mouth since no one wants to openly admit they were in charge of a first-class failure. Reading articles from the trade press and attending seminars tends to give the impression that the vast majority of IT implementations are a success, and even assuming someone did offer to admit via these forums that their project failed, who would want to listen? In general people want to learn by listening to, and ultimately emulating, a success story. But we also learn by our mistakes and admitting them openly allows others to follow those who lead and avoid the pitfalls encountered on the way. The following simple rules will help to ensure successful project management—many of them have been learned the hard way!

- Clear roles, responsibilities and relationships are required
  - Each key decision should be clearly allocated to one person or team
- Ensure that the scope and objectives of the project are clear
  - These set the directions to and limits on the project manager
  - They also give higher management control over what is happening
  - Make sure the project boundaries are clearly defined at all stages
  - If any areas are to be excluded, make sure this is made known
  - Ensure the terms of reference are understood. e.g. speed or economy versus polish or comprehensiveness
  - Treat every proposed change with respect since errors may be introduced
- Estimate correctly
  - Estimate tasks at the lowest level possible since an estimate made up of many small elements is likely to be more accurate than one made from a few large estimates
  - Low-level estimates ease the task of monitoring progress
- Plan as often as possible
  - Ensure that an effective and correctly sequenced series of tasks is set
  - Use smaller tasks to motivate staff where necessary
  - Anticipate problems and hold-ups
- Plan realistically
  - Avoid setting false deadlines
- Build comprehensive product/module and work breakdown structures
  - Understand clearly how requirements will be met
  - Make sure the breakdown of each stage covers all the required work elements
- Communicate
  - Ensure that it is effective and continuous
  - Make sure everyone knows what is expected of them and when it is required
  - Ask for advice and contributions from all levels
  - Tell any bad news as well as good news

- Involve and empower people in the project
- Manage changes properly
  - Changing objectives, priorities and resources and reallocating tasks cannot be done effectively unless the current situation is clear
  - The impact of changes needs to be clearly understood and communicated
- Control the project
  - Measure the effort actually involved in tasks and compare with estimates
  - Use this feedback to ensure that the rest of the project is likely to be achieved
  - Manage the resources
  - Estimate requirements—even inaccurate estimates reduce the uncertainties in a project

7.7   Consultancy and external assistance

In Section 7.4 we discussed the makeup of the ideal project manager and noted that hiring an 'outside expert' has been shown to produce poor results. This is mainly due to one or more of the following factors:

- Internal persons soon learn what is required anyway
- It can sometimes take longer for an outsider to learn about the company (its products, processes and people) than vice versa
- The external expert may be an unknown quantity to the company—it takes time to build up trust, credibility, etc.
- Internal employees tend to take a back seat if outsiders are brought in. They do not feel involved and committed ('it's his job to install it') and will be the first to find fault when it becomes operational

These points are also valid if the project manager is a new company employee—perhaps the credibility and trust required has not yet been established. The chosen person must know the company, its people and products—ideally someone who has been there for at least three or four years.

Similar factors relate to the use of external consultants. Perhaps the implementation is expected to take an extended timescale and external resources will be required or, since EDM technology expertise may not yet exist to the required level within the company, externally experienced EDM specialists may need to be brought in to assist in ensuring a smooth implementation phase.

Whatever the reason, one important point should always be borne in mind: like going on a diet, no-one can implement EDM for you—it must be done by the organization's own employees. Outside advice, practical help and assistance can be sought and obtained in the form of consultancy but the ultimate responsibility for it should be taken by the company itself. This ensures that implementation responsibility is coupled to operational responsibility. Making external people responsible for an EDM system implementation has been shown to result in a poor installation.

Having said this it is still very useful to obtain external assistance and outside help should always at least be considered since, in general, the installation of EDM is an unfamiliar task to those company personnel involved.

The exact nature of the consultant's experience and the client company's requirements will determine where the consultant should be placed in the management structure for the project concerned. If consultancy assistance is required at the strategic level then perhaps a direct report/liaison with the steering committee chairperson is best.

However, the most effective and favoured situation is where the consultant works in conjunction with the project manager to provide the necessary industrial experience, resource, product knowledge, coordination, etc. to help the project manager achieve a successful implementation. At this level the consultant is hired to oversee the project and should not be expected to write procedures, draw flow diagrams, write descriptions, develop programs, etc. unless specifically requested by the company to do so. The consultant should only be called upon three or four days per month to review the situation with the company. More often, and the consultant tends to take over and can also prove to be very expensive. The consultant should, however, be available by phone between these days for information, routine contact, etc. During the consultant's visit time should be taken to:

- Meet with the steering committee, chairperson and project manager
- Meet the project team
- Liaise with all required
- Ask questions which force people to address the relevant issues
- Review progress and formulate recommendations
- Help people to focus on the correct priorities
- Provide guidance (act as a 'sounding board')
- Attend steering committee meetings

However, there may also be occasions, mainly due to a lack of resource, where consultancy help of a lower-level nature may be required to assist in specific task force operations where a more 'hands on' approach is required.

Whatever level of help is sought, competent and experienced consultants are required, professionals who know your industry sector and EDM technology. But where can such help be found and what kinds are available? As more and more organizations realize the benefits which can be obtained from EDM, the more need for appropriate consultancy and implementation services is generated. Many companies now provide these services, either as main revenue generators, or as part of their other normal activities. A number of the major suppliers are listed in Appendix A. They generally fall into four main categories:

- *EDM vendors* In view of their vertical market penetration, these vendors will have had the most experience in implementing advanced EDM solutions. However, this position is now being challenged by other newer players and therefore their relative share of the implementation services market is declining.
- *Major platform suppliers* Vendors such as IBM, DEC, Bull and ICL are gaining market share in EDM implementation services mainly due to their general drive into the services marketplace and systems integration in particular.
- *Systems integrators* These players are growing fastest in the consultancy and implementation services marketplace in line with the general move to SI-type business and the heavy integration aspects associated with EDM.
- *Management consultancies* Traditional consultancy companies such as Andersen Consulting, Coopers and Lybrand and PA Consulting Group have now moved into the operational side of technology consulting and through a mixture of practical experience and strategic alliances have already captured a reasonable share of the EDM implementation services marketplace.

Most of these sources, perhaps with the exception of the EDM vendors, will also provide a wide range of other services which, although more general in nature, can also be applied to EDM. These include requirements analysis, ITT/RFP preparation, cost/benefit analysis, tender evaluations, change management, education and training, solution design, and so on.

8

Practical considerations

8.1 Overall features and terminology

EDM is a productivity tool for the engineering and manufacturing industry—
an enterprise-wide facility to support product and service teams. It can be
used to manage all product-related information (including digitally-held data
files and database records) and is capable of controlling the entire product life-
cycle. Let us review what is meant here before moving on. The main elements
include:

- *The staged release of product information* The concept of concurrent or
  simultaneous engineering which was introduced in Chapter 1 has the effect
  of allowing the staged release of information relating to a product, rather
  than a more rigid sequential approach. EDM provides the ideal mechanism
  for controlling such data. This concept is shown in Fig. 8.1 where we can see
  the original sequential series of activities at the top, and the more flexible
  approach of parallel activities at the bottom.
- *The correct use of configuration management* At each stage in the product
  release process, and on an ongoing as-required basis, EDM can be used to
  generate a 'snapshot' or change history of the product-related design/pro-
  duction or service data at any particular time. i.e. a configuration record.
  We have already seen how important such records are with regard to traceability and quality management.

  It is also important at this stage to appreciate the difference between
  product structure management which is the simple maintenance of a hierarchical product structure and configuration management which encom-
  passes design releases, changes, secure file areas and associated file
  references, etc. Product structure management is therefore only one of a
  number of different items used to provide a full configuration management
  environment.
- *Change management procedures* Any company which is going to release
  product information and maintain configuration records on an ongoing
  basis must adopt a suitable change control process. EDM will provide a

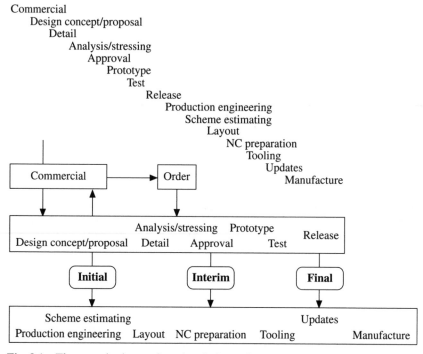

Fig. 8.1 The staged release of product information (*courtesy of PA Consulting Group*)

structured environment to log the change itself, move it through a predefined sequence of events, obtain approval where necessary and keep a note of the change history.

● *The use of suitable approval processes* In addition to the approvals required to pass a change notice from one stage to another, EDM systems allow other approval mechanisms to be used, such as information access approval mechanisms, project-related approvals, and so on.

These elements need to be controlled and managed over a wide range of current and future business requirements and technologies. For example:

1 *Combinations of many products and projects* Since many companies are now adopting very different product strategies and working practices than they used in the past, an EDM system has to be able to cater for these requirements in a comprehensive and flexible manner.

2 *Geographically distributed* This includes the trend to operate over a number of sites, often spread over a wide geographical area to take advantage of resources, distribution or development requirements.

3 *Heterogeneous platforms* This is simply 'computer speak' outlining the

need for EDM systems to operate over different computer hardware 'platforms' (i.e. processors and peripherals) in a simultaneous manner. This reflects the trend towards open system environments and is directly opposed to a homogeneous environment where only one supplier's system is dominant on the site.

4 *Heterogeneous applications* In a similar manner to heterogeneous platforms, heterogeneous applications are software programs (for example 2D CAD) which are required to run on any number of different hardware platforms.

5 *Over a network of subcontractors* A further example of changing business practices and environments, where partnership sourcing and the use of subcontractors may necessitate the use of EDM over a wider user-base with remote access capabilities, and will often also relate to the geographically distributed environments we have just outlined.

So how can we recognize an EDM system if we see one? What form does it take and what kinds of people use it? The answers to these types of question are sometimes not so clear-cut. This is because the scope of systems is wide and varied: they range from the very small to the very large, they cater for many different user needs and can be heavily customized. However most EDM systems can be considered as belonging to one of the following four types (refer to Fig. 2.8 to help position them in a company environment):

- *Single workstation systems* Perhaps on a small site where most of the requirements can be fulfilled by the use of a single workstation or terminal. The scaleability of the chosen solution is important here to cater for any future expansion.
- *Workgroup/department-based systems* This could relate to the provision of EDM functionality for a project development team or department (for example a drawing/design office) in a specific, vertical type of role. Again, expansion potential should be considered.
- *Product-related systems* These types of environment relate to the use of EDM on a product-by-product basis, where the system would need to be used to control an important contract or product before being expanded to other products, or for general use, at some later stage.
- *Enterprise-wide systems* These types of system have either grown into, or been specifically purchased as part of, a major investment in order to provide a strategic benefit to the organization concerned. Such systems are strategically important and many contain elements of on-line transaction processing (OLTP) and fault-tolerant system technology to guarantee the uptime of the solution concerned.

In a similar manner to the scope of the systems themselves, the terminology used can initially be confusing and very often different terms are used for like-meaning items. The following list explains some of the most commonly used EDM jargon. Each system vendor uses their own variants of these—and a few more besides!

- *Check-in and check-out* This refers to the act of obtaining a data file from some form of secure area (i.e. checking the file out) or returning the file to the same kind of area (checking the file in). It is generally used when files need to be updated, perhaps as a result of an ECR, but files can also be checked out for copy purposes. Read-only access does not normally require a formal check-out process, but a record of the file having been accessed (or 'touched') should always be kept.
- *Electronic vault* This is the term often used to describe the secure area referred to in the above term. It is the master storage area for the associated files which the EDM database manages and can physically be of a distributed nature. Logically, however, it appears to the user as one large secure area where all the issued data files are located. Note that these can be of a pre-production and/or production nature depending on the system and desired method of working. Files with a 'working/draft' or 'to check' status will usually be held outside the vault, but again in some form of secure area, until the appropriate level of authority is gained, normally via the change control system.
- *Revision and version* These two terms are generally used to describe the release number of the item under consideration and both relate to the normal meaning of the words as used in manual updating procedures. For example, a pre-production version control system may number individual releases as a, b, c, d, . . . while a production system may use 1, 2, 3, 4, ...
- *Metadata* This is the term which, as we have seen, is often used to describe information about information. In the context of EDM it could be used to refer to the data held in the EDM database which in turn points to the associated data files themselves. This concept is shown in Fig. 8.2.
- *Object* An object in an EDM system is usually taken to mean a particular document, file or part/item/sub-assembly record which can be represented individually on the operator's screen. It is a generic term and used in preference to FEM file, CAD file, DTP file, etc. The term object is also used in RDBMS technology and object-oriented programming systems (OOPS) such as C++.
- *Encapsulation* This describes the way many EDM systems control access to the individual files and objects as presented on the user's screen. Each object would have certain information and rules (or scripts) which surround it. When the object was 'clicked' or 'touched' by the user (perhaps via mouse or keyboard control), these rules would be read and actioned, thus easing the user's task considerably. For example, a 2D AutoCAD file may be

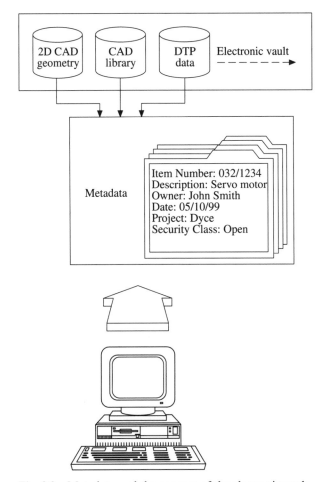

Fig. 8.2   Metadata and the concept of the electronic vault

encapsulated with information relating to the version of AutoCAD to run when the file was touched, the machine where AutoCAD can be found and any associated parameter files or viewing variables which may be needed.

- *Browser*   This is simply a term used to describe a method of accessing data within an EDM system on a read-only basis via a simple and convenient user interface. i.e. browsing through the database for information.
- *Markup/redlining*   These are similar terms to describe the act of obtaining access to a data file via the EDM system and 'overlaying' a data mask onto the file which can optionally be stored separately. This could, for example, suggest changes to be made to the object concerned as part of an ECR to be circulated around for approval, and is analogous to using a red pencil to highlight areas on a traditional paper copy.

- *Electronic sign-off*   This relates to the approval of a document or object via the computer system, thus obviating the need to maintain a signed copy. Although now widely used and accepted in many environments, electronic sign-off still creates some concern among certain system users. Many companies (and individuals with personal legal responsibility) still want a paper copy and wet ink signature to replace or back up any electronic sign-off. It is still not seen as sufficiently foolproof by many regulatory authorities and by many individuals who stand to 'carry the can' if something goes wrong.

Before we leave this section and consider some of the more specific system features in more detail, let us consider the types of user who can benefit from EDM system technology. As we have seen, the type of information which can be accessed and the potential of the database and files stored in an EDM system is vast—almost everyone can therefore benefit. System users include designers, engineers, manufacturing specialists, managers and executives, administrators, etc. They generally fall into three main categories: end-users that create or modify information, managers responsible for projects and organizations and administrators (both EDM and organizational):

- *End users*   End users can include almost everyone in a given enterprise. In general, the EDM system appears to these users in one of two ways: (a) additional screen menus and/or functions within the normal application environment, or (b) a 'shell' over their existing software applications which automates many of the access and modification functions. e.g. data retrieval, storage, backup, etc. Typical end users include:
  - CAD/CAM operators, designers and users
  - NC programmers and operators
  - Manufacturing engineers
  - Design engineers
  - Configuration control staff
  - Shop floor personnel
  - Purchasing staff
  - Marketing and sales personnel
  - Service and support staff
  - Work study/methods engineers
- *Managers*   These include persons responsible for project and product planning, company departmental control, task assignment, tracking and reporting, budget maintenance and general review and approval type operations. For example:
  - DO managers
  - Engineering managers
  - Supervisors

- Quality managers
- Project managers
- *Administrators* Organizational administrators include people responsible for the daily operation of individual business enterprises, for example the authorization of work, part number registration, data distribution and configuration control functions. They generally report to departmental managers. The role of the EDM administrator is more akin to the database administration function outlined in Chapter 4, in that he/she/they (depending on the size of the implementation) manage the implementation and maintenance of the EDM system itself. The EDM administrator's overall knowledge of the system will far exceed that required by individual users and managers. They require a balance of technical, personal and business-related skills since their ability to encourage and support EDM users can have a significant effect on the success of the implementation and its eventual use. Typical administrators include:
- Systems/data analysts
- Filing clerks
- Development staff
- Operators
- IT management
- Database administrators

8.2 System features and procedures

In Chapter 2 we looked at the key elements of EDM and their associated functionality. This section will build on this information. We will revisit some of the elements, expand on others, and look at these features and procedures in more detail. This will allow us to consider EDM in a different manner—to explain its operation from a more practical viewpoint rather than a purely theoretical perspective.

User interfaces

The user interface of an EDM system is a key factor in how productive the tool will be and as such is a vital component in establishing the eventual success of the overall solution within the company concerned. Modern computer systems require user interfaces that are graphical, intuitive and tailorable if they are to be liked and accepted, and EDM systems are no exception. In fact, ease of use is probably even more important with EDM systems since they are used by a diverse set of people with various computer skill levels. There are two main elements to be considered here: the 'look and feel' aspects which determine how the user interacts with the system, and the 'man–machine interface' which is the process invoked within the 'look and feel' to perform a specific function.

EDM user interface styles are wide and varied, such as command-line, forms-based and menu-driven, with each individual user being able to 'see' only those options they have been given prior authorization to access. Interaction with EDM systems can generally take place on both graphic and alphanumeric terminals. Alphanumeric terminal access, because of its traditional and more widespread nature, is heavily used but can of course only access the metadata or part-related data from the EDM database itself. Any attempt to access graphics data should result in the appropriate error message being displayed. Graphics terminals on the other hand, because of their ability to provide a graphical user interface (GUI), can not only access the metadata and display and run the individual applications which generated the data (e.g. CAD, DTP, etc.), but can also enable easier user interaction and data display via the use of appropriate window technology. Because of the nature of EDM, the devices used to access the system require to be of a heterogeneous nature, for example, VTxxx, IBM3270, PCs, Apple Macs, X-terminals and workstations.

EDM user interfaces are now heavily GUI-based, using standards such as OSF/Motif, as more and more graphics-related technology becomes available via networked PCs, X-terminals and workstations. Most suppliers will provide both alphanumeric and GUI-based user interfaces and it is important that both styles maintain as much commonality and compatibility as possible to allow users to work in a consistent and effective manner.

General user interface considerations also apply, for example the number of steps required to complete a specific function, the intuitive operation of the interface to accommodate 'casual' users and the provision of on-line help information. Two typical user screens are shown in Fig. 8.3: (a) showing a general screen of image files and menus, and (b) showing the results of various enquiries overlaying the associated graphical information.

File access control

The ability to enable a user to access a file from a secure area or data vault, or to change the status of a particular file or object from say 'checked' to 'approved', may (and indeed should) appear very easy. However, such interaction can only be initiated by a user who already has the necessary authority. The alterations to the file's access settings are made within the database, but under the strict control of the EDM system itself, thus offering a superior level of control over file access. EDM should replace the traditional computer concepts of access rights like 'read-write' with real-world concepts which are more meaningful to engineers, such as checking, approving and release. These control features are exercised by authorized users through fairly 'transparent', or perhaps automatic, operations involving status. Of course, someone has to administer such access control settings and this is typically done by creating

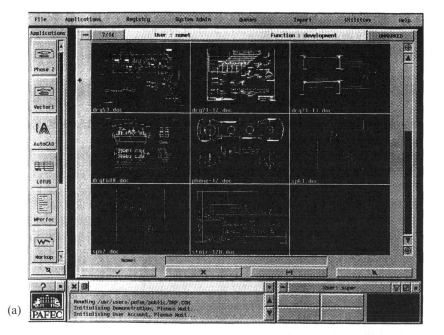

(a)

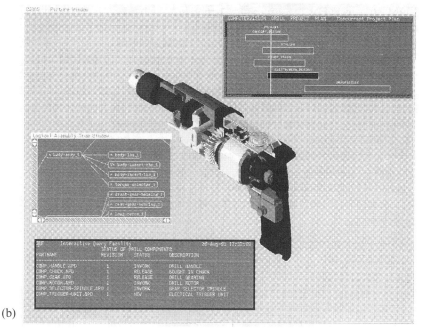

(b)

Fig. 8.3   GUI-based user interface screen examples ((a) *courtesy of PAFEC Ltd;* (b) *courtesy of Computervision*)

suitable lists and storing them within the EDM system. The lists provide the necessary links between different users or user classes, and their associated access privileges.

Data transportation and translation

The need for transparent access to the type of data outlined above, where the access rights are managed by the system, not the user, is just one aspect of the wider concept of transporting data around the EDM system and having it in the right place at the right time.

EDM systems effectively isolate the user from having to know where a specific data object resides and how to access or move it. The necessary network and operating system commands are hidden in a similar manner to the access rights. For example, a user could simply locate an object via standard search facilities and, when selected, the EDM system would move the object data between the secure storage area (vault) and the user's application workspace, and vice versa, i.e. the check-in/check-out process. Such operations should be transparent to the user and operate over a heterogeneous hardware and software environment.

Data translation relates to the specific need to interchange product data between different application systems which may form part of the overall EDM solution. For example, files being moved between such applications must often be translated from one application's format into the others before use, such as between dissimilar CAD and CAM systems.

Translation functions can be manually initiated or performed automatically by many EDM systems as part of their configuration capabilities. The above example may require the CAD file to be fed through an IGES translator before being successfully recognized in the CAM system—all such operations having been configured into the system by the EDM administrator for transparent operation by the user. The translators themselves can be customized or standards-based, and procured from many sources (e.g. application vendors, EDM vendors or specialist suppliers), and may indeed be subject to control by the same EDM system which uses them!

In summary then, data transport and translation within an EDM system includes:

- Copying or moving data files as required between users, applications and (if applicable) different systems and sites. This should assist in maximizing the use of any 'legacy' data from older technology solutions.
- Ensuring the ability of the EDM system to know the nature of any translations required, for example any different file formats or types which may need to be changed automatically when moving from one application or workstation type to another.
- Operation over Local and Wide Area Networks.

- The use of industry standard networking services.
- The ability to handle and make the correct use of the different file types required within the system, for example native application files, PDES/STEP, CGM, TIFF, IGES.
- The provision of conversion software. This may be standard (e.g. IGES), user developed, or commercially available translators.

Design release / life-cycle management

Many EDM systems use these, or similar, terminologies to define the stages in accessing data and moving the engineering document or object through its life-cycle. This is usually performed by controlling the status of the document, via the standard system functionality, from 'draft' through to 'released' on an iterative cycle as new releases are made. We discussed this sequence of operations in Chapter 2. Controlling the status of individual documents, or groups of documents, is a user service which involves the system transparently accessing the product data vault via suitable check-out and check-in facilities and provides a controlled and secure environment for the storage of all the data sets managed by the EDM system. This could include:

- The product and engineering-related database itself, i.e. the metadata. This will include the change control and product structure information.
- Any other associated administrative information.
- The references and actual data files relating to the database (metadata) itself.
- Any external data references to perhaps microfiche or paper-based information.

It is therefore considered to be the core function of an EDM system, and if utilized on its own (as an EDM system in its own right or part of a larger system), provides the basic file management facilities needed to support the additional aspects of change control and configuration management.

The definition of a new item or part record in an EDM system can be initiated via two main routes. Initially, the part information could be defined in the EDM system database itself: its part number, description, remarks, etc.—the metadata. This could be followed by the generation and linking of the associated data files to this metadata. Conversely, the initial generation of the data files via the application software itself could be followed by the prompting and subsequent generation of the metadata from the filing routines of the application (providing the application can be suitably configured).

The steps involved in the generation of such data are probably best explained by considering a simple example of a designer or draughtsman creating a 2D drawing on a CAD system. They may work on the drawing for some considerable time and may choose to operate from a local disk, at least initially. This could be located on the workstation being used, or maybe a local office file server—whatever the case, initially the file is considered to be outside an EDM framework. At some point, once the drawing takes on some reasonable form or when the system or company concerned dictates, the drawing needs to be filed under the control of the EDM system. The operator would then check it into the electronic vault of the EDM system via the appropriate filing routine. At this stage the file would have a 'draft' status flag set. The operator could then work from this environment, by checking out and checking in the drawing to the draft status section of the vault, until the drawing was complete and ready for checking. At this point the drawing could be filed as a 'to check' status and the appropriate message would be sent to the person authorized as the 'checker'. The checker would then check the drawing out, approve it and check it back in with the appropriate status, and so on until the drawing became a fully released version. Once a file, document or data object is checked in (at any status level) full EDM system security, access control and change control (if available) will be in force. If and when full change control is used, and an element is modified and approved, it will be checked in to replace the previously controlled design. In order to provide configuration management and control, the previous version will be saved and not overwritten.

A check-out operation is performed when controlled (i.e. authorized) access to information is required, such as:

- A simple viewing operation of a particular object
- Use of an object for copy purposes
- An edit operation without the need for change approval (i.e. a 'working' file)
- Access to a file for redlining, markup or comments as per the change control cycle
- When modifications or changes are required, perhaps as part of the ECR/ECN cycle (note that the system should lock the information in this case and prevent other users from concurrently modifying the data, although read-only access is required)

A typical document check out screen is shown later in Fig. 8.7.

The design release functions of EDM systems also define all the system users together with their associated authorization levels, and the release or promotion levels required for use with the change control system. All user authorizations are checked before any design release management function is completed and, since each change process may require different release levels and/or checkers and approvers, the project manager or system administrator must be able to select those users required to complete these tasks in a simple and flexible manner.

Finally, EDM systems provide the capability to link multiple objects and data sets together via design release management. Such data objects may be associated with the same part or assembly, for example an engineering drawing and its associated specification and test schedule. If these were linked together, the EDM system would be required to automatically track and maintain proper version compatibility as each separate member of the set was modified or changed. By using such facilities, the concurrency of the data files is maintained and the accuracy of the entire database improved considerably.

Data searches and part classification

EDM systems, by their very nature, have to provide comprehensive data search and retrieval facilities. Part of this functionality requires certain product data 'attributes' to be assigned so that similar items can be related in a way that eases their retrieval from the database. In this way the system can be used to search for existing documents, data, parts, assemblies and standard items which will maximize their re-use, thus increasing the potential of product standardization and reducing product costs. The actual data fields from the meta-database which could be used as search criteria will of course depend on the system concerned and the way it is configured, but the following are typical examples: part/item number, function code, project, status, classification code (see below), description, release date. These data fields, and the attributes they contain, can be assigned a default value by the system or can be user assigned. The ability to perform such searches is of course directly related to the general search capabilities of the RDBMS itself, which is employed to hold and manage the meta-database structure of the EDM system. However, the EDM system itself is required to extend this functionality and provide specific menus and user functions to classify and retrieve relevant information. These menus and screens should allow the system user to specify ranges and logical combinations of the data fields outlined above for both screen-based output and report generation, and can very often be stored to allow the user to re-query the system without having to re-enter the query from scratch.

The classification code mentioned earlier is generally considered to be part of a more structured form of 'part classification and coding' and could be a direct part of the EDM system functionality (providing class structures and codes), or a link to some form of external part classification system (to provide a standard classification scheme). The best known, but dated, examples here are the Brisch and Opitz systems. A classification and coding system is not only used for data searching and retrieval, it is also a fundamental prerequisite for the adoption of group technology or cellular manufacturing, a technique which can provide significant competitive advantage in the area of manufacturing flexibility.

A typical data enquiry/search input screen is shown in Fig. 8.4.

```
Part Find                EMCS-X ENGINEERING MANAGEMENT

1. Keyword        =  ~  FULLNUT        2. Alias  % :
   Type           =  ~
   Number Series  %  :  81
   Project        =  :
   Description    %  :  HXHD%BRASS%
   Part/Document  =  :
   WIP/Issue      =  :                        Number found : 5


  Part number  TYP UOM  Description                      User Group  S

  810500       CAT  EA   M5  HXHD FULLNUT          BRASS  standard
  810510       CAT  EA   M6  HXHD FULLNUT          BRASS  standard    I
  810550       CAT  EA   M12 HXHD FULLNUT          BRASS  standard    I
  810551       CAT  EA   M30 HXHD FULLNUTS-BRASS    PLT   standard    I
  810720       CAT  EA   M10 HXHD FULLNUT          BRASS  standard    I



    select(1)      query(2)      properties(3)    wip(4)    enquire(5)>
```

Fig. 8.4 A typical data enquiry/search input screen (*courtesy of MPSL*)

Change management

We have already discussed how important the change management process is in today's business environments—EDM systems provide the flexibility required to enable such processes to be configured around site-specific business rules. We have also seen how access control and design release management can be used in conjunction with these change control facilities to provide the framework necessary for successful change management. This section will therefore review and summarize the main elements of EDM change management functionality and work through an actual example to explain its operation in more detail.

Change management functionality typically includes the ability to:

- Ensure changes occur according to the appropriate predefined site-specific procedures
- Maintain an audit trail or change history for every relevant item
- Specify approval mechanisms
- Notify actions (perhaps for check/approval) via E-mail
- Electronically 'sign-off' approved documents
- Modify and add product data via ECR/ECNs
- Assist in the control and timing of changes into already released or baselined data
- Synchronize changes, thus ensuring that a change to any related data item does not invalidate the associated data relationships

● Maintain appropriate data/object status codes during the change management cycle

Let us consider these in a simple example. Each step in the following example is accompanied by an actual screen shot from a representative EDM system. In this case the example material is reproduced with the kind permission of Sherpa Corporation.

Step 1

The change process is normally triggered by the creation of an Engineering Change Request (ECR). This could occur as a result of a customer complaint, a market requirement or perhaps internally via any part of the design/DO/manufacturing/service cycle. The person who raises the ECR needs to define the parts, objects, assemblies, etc. to be modified and provide as much information on the reason and nature of the changes as possible. The ECR may require the provision of additional files or maybe paper-based information and/or certain files to be redlined to detail the area of change. The creation of an ECR is shown in Fig. 8.5. Once complete it can be E-mailed to the appropriate section(s) or person(s) for approval, depending on the specific procedures adopted by the company concerned.

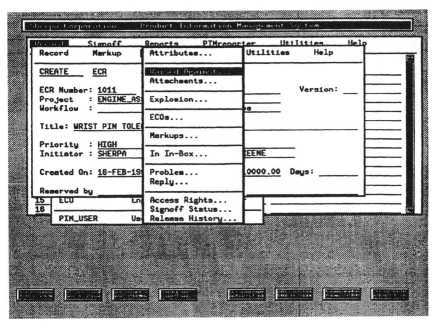

Fig. 8.5 ECR creation example (*courtesy of Sherpa Corporation*)

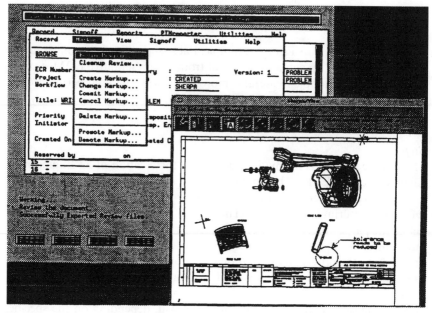

Fig. 8.6 ECR review/check operation (*courtesy of Sherpa Corporation*)

Step 2

The ECR is reviewed/checked and approved (via access to the electronic vault) through the required sequence and in line with the status defined in Chapter 2. This may necessitate a serial review and electronic sign-off process by those members of the change committee, or perhaps done jointly at a change control meeting. Whatever method and degree of refinement is chosen, the ECR data will need to be accessed and checked. This is shown in Fig. 8.6. Here we can see the ECR data in one window while the assembly is shown (in marked-up form) in a separate window running the CAD application to view the relevant file.

Step 3

Once formal approval is given, work can begin on modifying the object(s) in line with the requirements of the ECR. The first step would be to withdraw it 'for update' from the electronic vault. This is shown in Fig. 8.7. The user would then be able to select the item(s) on the ECR for update and the EDM system would copy the files into the user's work environment and run the application associated with the file itself (i.e. operate on the file's encapsulated data). In the example shown in Fig. 8.8, a 3D CAD application is initiated.

Other users who access the file for viewing and copying would receive an appropriate message and be allowed to continue; a user requesting the same file for update would be locked-out from doing so.

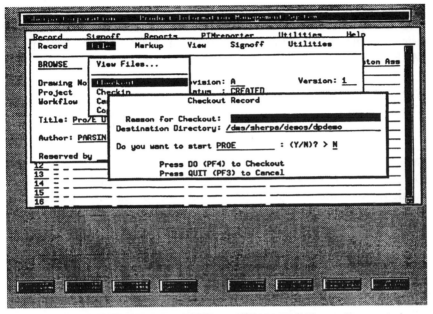

Fig. 8.7 A typical document check-out screen (*courtesy of Sherpa Corporation*)

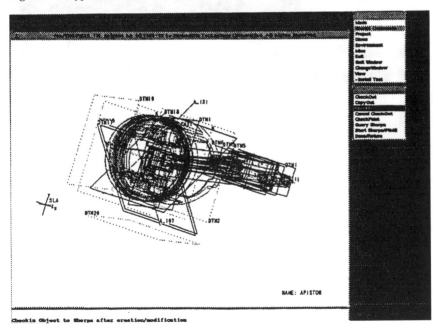

Fig. 8.8 File modification—example CAD application (*courtesy of Sherpa Corporation and Parametric Technology Corporation*)

It should also be mentioned that, although our discussions to date have centred on objects and data files, the parts/items list data can also be changed using a similar route and fed to and from an external source—perhaps an MRPII system. This is discussed in more detail later.

Step 4

The actual operation of updating the relevant information is effected. In our case the assembly is opened and ready to be modified as shown in Fig. 8.8.

Step 5

Once updated, the file(s) are returned to the control of the EDM system's electronic vault via normal exit operations from the application concerned. The changes can then be checked and approved as necessary in a similar manner to those outlined in step 2. This can either be performed electronically via a graphics terminal to view the changes and the ECR simultaneously using window technology, or via a simple alphanumeric terminal to update the ECR itself after suitable verification of the changes, perhaps via paper-based copies. Figure 8.9 shows the actions of a checker signing-off an ECR electronically.

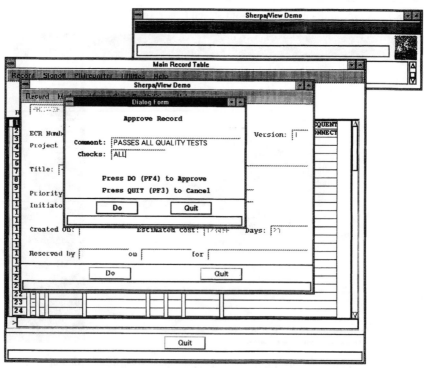

Fig. 8.9   ECR sign-off/approval (*courtesy of Sherpa Corporation*)

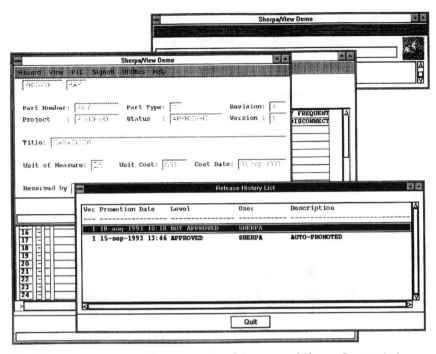

Fig. 8.10   Approval/sign-off signature record (*courtesy of Sherpa Corporation*)

Step 6
A record of the transactions is kept throughout the entire course of the change control process. Figure 8.10 shows a record of the approval signatures held with the other part and assembly metadata.

Step 7
Once final approval is given the data is checked back into the electronic vault with its version number incremented as required. This operation is usually prompted by the system and checked on input, but can be manually overwritten if for example a step-change in the revision/version numbers was required. (e.g. . . ., 4, 5, 6, 20, 21, 22, . . .). A typical screen is shown in Fig. 8.11.

Step 8
Finally, an Engineering Change Note (ECN) is usually generated to communicate to all relevant users that the ECR has been effected and the necessary changes made. Again this could be manually or electronically circulated.

Product structures
The ability to manage product structures in an EDM environment is fundamental to the correct adoption of configuration management (or CMII as it

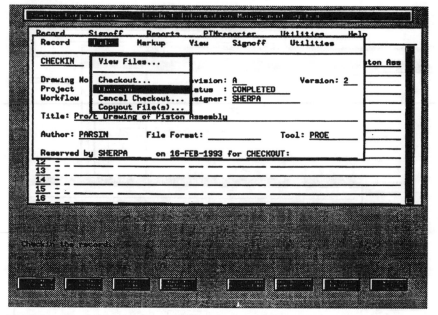

Fig. 8.11 Check-in operation example (*courtesy of Sherpa Corporation*)

was introduced in Chapter 2). The maintenance of comprehensive configuration management and traceability records lies in the ability of the EDM system to maintain suitable product structure records—to create and maintain parts list (PL) or bill of materials (BOM) information relating to the file data and metadata held within the system itself. This basic product structure will generally be enhanced to store the variants, versions and the date or version effectivity criteria required.

The search and classification type facilities previously explained are enhanced considerably by the use of product structures, allowing users to easily find all the information related to similar parts, assemblies, items lists and BOMs. The generation of the product structure itself can be direct, with the user interacting with the EDM system to create and modify BOMs, associate information, access data and, eventually, to create reports. This BOM/PL structure could then be exported to, say, MRP. Alternatively, if a product structure already exists in MRP, the EDM system should be capable of accepting this data through suitably transparent and efficient access methods for subsequent modification and edit purposes.

A BOM is often considered as a form of 'structured' parts list that shows what material goes into a product. Much has been written on the subject of BOM structuring—the logic of it is important to many different company functions which may visualize it and use it very differently, for example engineering (dependency calculations, where-used information), cost accounting

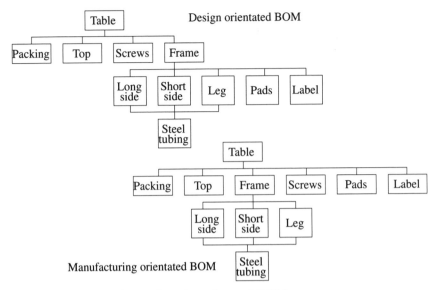

Fig. 8.12   Design and manufacturing orientated BOMs

(costing simulations, cost roll-ups), production/manufacturing (phantoms, MRP) and field service (spares availability, actual structure representations). It is important that, although these different views of the BOM are possible to obtain, they are derived from a single product structure definition in order to maintain structure integrity and minimize the effects of the data transcription errors which would inevitably occur if multiple BOMs were held. EDM product structure management facilities should also include the ability to define alternative and optional parts—these again reduce the number of different BOM structures which need to be held and allow more effective management to be exercised over those that are really needed. The main differences are encountered between 'as-designed' and 'as-built' BOMs. A simple example of these two views is shown in Fig. 8.12 which illustrates different views of an office table made from a tubular steel frame and wooden top.

In this specific case we can see that the designer sees the frame as being a major sub-assembly of the table and wishes the BOM to reflect this. The production engineer, on the other hand, sees the frame sub-assembly as being the welded and painted structure only, which may be produced and stocked separately. The frame, top, screws, pads, etc. can then be brought together during the final assembly operation.

Other differences between design and manufacturing BOMs can occur for many other reasons. For example, 'phantoms' require to be correctly set up in a manufacturing BOM to ensure that they are treated correctly in the MRPII system (if applicable). There may be some items on a BOM that engineers do

Level 012345	Item number	Description	Qty	UOM
0	-----	Table	1	ea
1	-----	Packing	1	roll
1	-----	Frame	1	ea
2	-----	Long Side	2	m
2	-----	Short Side	2	m
2	-----	Leg	4	m
2	-----	Pads	4	ea
2	-----	Label	1	ea
3	-----	Steel tubing	X	m
1	-----	Top	1	ea
1	-----	Screws	8	ea

Fig. 8.13   An indented PL/BOM listing

not designate by a specific part number—e.g. a casting that can be plated or painted in different finishes. This may only have one part number for engineering purposes but production requires separate part numbers for each variation in order to correctly plan requirements. BOMs may also use 'non-engineering' numbers to assist in scheduling requirements. For example, an assembly may have brackets, nuts, bolts, etc. used to fasten the assembly together. Rather than have all these items appear, an imaginary assembly or 'kit' number may be generated to cover all of the necessary parts. The 'kit' number is not a real assembly—it is simply used to identify an imaginary bag of parts to make the final product.

BOMs have traditionally been viewed via an 'indented' listing. Figure 8.13 shows an example for the design view of the table we discussed earlier. Although each level is represented on the BOM to establish the parent/child relationship, it is much more difficult to visualize the structure in this form. EDM systems now generally provide an option to graphically display the product structures that show multiple assembly levels in a much more user-friendly format. Two different examples are shown in Fig. 8.14. This type of display greatly increases the user's understanding of the structure and the associated component relationships, and enables easy browsing of the entire product structure.

'Effectivity' is an important attribute of any BOM or product structure. This is used to determine which document/item/part revisions should be used when a product is assembled, i.e. which revisions are 'effective' at the time of assembly, or used historically to determine the revisions of items/parts used in a particular revision of a product, assembly or sub-assembly. Different effectivity schemes should also be provided—these would be specified during the

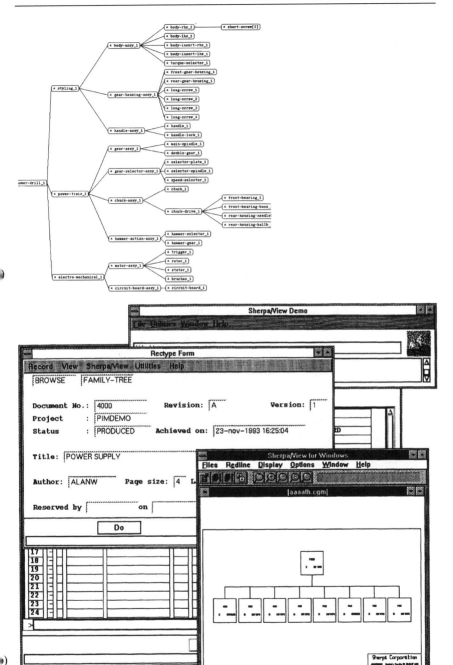

Fig. 8.14 Assembly product structures—graphics-based display examples ((a) *courtesy of Computervision*; (b) *courtesy of Sherpa Corporation*)

change control process. The change control process should work in conjunction with product structure management to ensure that effectivities are properly maintained. The three most common effectivity schemes are:

1 *Serial number*   This specifies which serial number of each major assembly or each end-item configuration is required and is often found in defence, aircraft and Government subcontractor environments, especially at the higher (master schedule) levels in a BOM.

   Serial number tracking often cascades downwards from the major assembly level in a BOM, but can become a cumbersome method and is generally not used unless absolutely necessary. Product configuration change control is effectively accomplished by using part/item numbers to define the configurations involved, and serial numbers (or dates, or a combination of both) to control change effectiveness. The choice of serial number or date is determined by the importance of the change itself and the indenture level at which it occurs. Changes which affect the interchangeability of serialized end-item configurations are best controlled by serial number.

2 *Date*   Date effectivity is often used at the lower indenture levels in a BOM (i.e. below the master schedule level), and in its simplest form specifies a date when a change becomes effective. It is often used in conjunction with the other methods as mentioned above. Changes that affect part number interchangeability at lower levels and do not affect end-item interchangeability are usually best controlled by simple date effectivity within the BOM of the unchanged parent part number.

3 *Lot/batch*   This is a further option to incorporate the change in the next batch or lot of goods being used or manufactured. It helps in some way to using existing stock and is particularly useful where an 'elective' change allows the opportunity to phase in a series of changes at one time.

Two other aspects of product structure management should also be briefly mentioned:

- *Attributes*   EDM systems should provide the ability to attach attributes to individual parts and to part/part relationships either for specific revisions or globally. These attributes, when tied to the part/document itself, can be used to provide more information than the basic fields that are provided as standard and are sometimes precoded (as for example field1, field2, field3, etc.), or are entirely user-definable. They are often used to assist in searching and locating already existing data, as we have seen. The attributes which can be attached between part relationships can be used to define unique information for each link concerned, for example to define the remarks data relating to the use of individual parts on different BOMs, or to define x, y and z coordinate information for each occurrence of a component on a PCB. These examples are shown in Fig. 8.15.

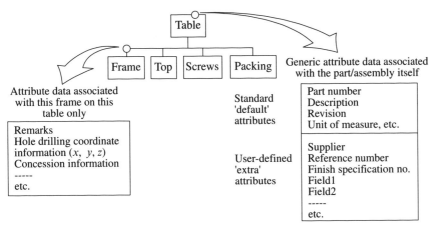

Fig. 8.15 EDM attribute data

- *Reports* A wide range of reports are usually provided to enable the often complex relationships of product structures to be visualized and interrogated. Typical examples include:
  - Indented BOMs (similar to the indented enquiry as shown in Fig. 8.13)
  - Assembly structure reports
  - Where used parts reports to many levels
  - Revision history reports

Messages and communications

Communication is the essence of an EDM system—it is communication, in any of its forms (for example files, application data, administration information, messages, images, etc.), which creates a concurrent engineering environment. All types of EDM users require to be notified of changes or information which may affect the data they are using or their work environments, and they also need to be able to send and receive messages relating to their day-to-day operations.

In response to this need, many EDM system suppliers now provide either their own electronic mail (E-mail) facilities, to inform users of the release of a new change request or of the approval of a new revision for example, or utilize the E-mail application which may already be resident on the host system's environment. Figure 8.16 shows an example EDM E-mail message screen.

Project management

As previously mentioned in Chapter 2, many EDM systems are now beginning to provide easily used 'hooks' to external project management software modules and/or optional system modules to control projects or 'programmes'

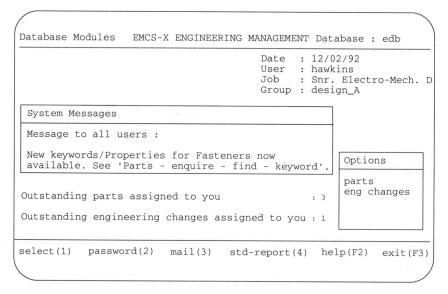

```
Database Modules    EMCS-X ENGINEERING MANAGEMENT Database : edb

                              Date   : 12/02/92
                              User   : hawkins
                              Job    : Snr. Electro-Mech. D
                              Group  : design_A

  System Messages

  Message to all users :

  New keywords/Properties for Fasteners now      Options
  available. See 'Parts - enquire - find - keyword'.
                                                 parts
  Outstanding parts assigned to you        : 3  eng changes

  Outstanding engineering changes assigned to you : 1


  select(1)   password(2)   mail(3)   std-report(4)   help(F2)   exit(F3)
```

Fig. 8.16   Example EDM E-mail message screen (*courtesy of MSPL*)

via a set of suitably integrated system functions (as opposed to the more traditional approach of bespoke interfacing to a standard project management application). The exact definition of a project or programme varies from system to system, but reflects the way many companies now operate in task-driven operations and environments. These could relate to individual tasks, through to the definition of a large order of a standard product as a project requiring many related tasks, and on to the control of a large one-off project, perhaps the building of a unique piece of plant or a construction project where specific BOMs and change requests were required to be defined and controlled.

Traditional project management tools such as Gantt charts and critical path analysis have been available as software applications for many years now, but EDM systems enhance these by providing tracking and status information down to task level, as well as change control and related product structure data. EDM systems provide a focal point for product planning and build a management knowledge-base, which offers technical advantages over the traditional segregated approaches to project management. Many EDM systems' project management modules are still fairly basic and are restricted to the creation and modification of work breakdown structures (WBS). These define major tasks, subtasks, and detailed tasks, with associated resources and expenditure profiles. This ensures that individual documents and work packages are coordinated throughout the operation. As the engineering process is refined to make it more responsive to market pressures, the patterns and rules governing workflow can be adjusted to allow these process changes. It is expected that this functionality will be further enhanced as market pressure continues and more extensive capabilities are demanded by users.

Imaging and control of raster data

Our discussions thus far have mainly concentrated on the control of application generated data files or documents, but it must also be remembered that EDM systems can be used to control any form of digital data. One of the fastest growing areas of digital technology over the recent years has been scanning and image processing, generally known as document image processing (DIP). For example, many companies have to maintain large volumes of data either by law or because the project contract has requested it. Such data, typically sales orders, purchase contracts, invoices and statements, etc., usually tie up large amounts of physical space. Scanning or electronically archiving such data can be a much more cost-effective way of dealing with it.

Document management, rather than EDM, is the name usually given to the practical use of DIP to provide the required business benefits. However, the concept of EDM can include the main aspects of document management— the image file is stored in exactly the same way as any other data file and can be retrieved and viewed under the same access control mechanisms. Additional image services include capabilities for creating (actual data capture via scanning or transfer from CAD), storing and distributing images via optical disks, annotating images via redlining and markup options and other general raster editing capabilities. Like any other EDM-related application, most EDM vendors provide these image services by integrating an image application supplied by a third party, although the control over the data files produced is a standard part of the EDM system itself. Many EDM suppliers enhance the functionality of the image software since the viewing, markup and annotation functions are EDM-specific and are especially useful in the change control and review process.

System administration

The role of the EDM administrator is a key one and has already been outlined in Section 8.1. The administrator provides and configures the entire range of system management functions, such as:

- Defining and maintaining the metadata
- Defining and controlling associated file references
- Controlling user authorizations and accesses
- Managing data archives and system backups
- Maintaining the distribution of data over the system

In order to perform these functions in a defined and controlled manner, EDM systems provide a number of utility features to ease the task of system administration. Much of this type of work goes on behind the scenes to minimize any end-user disruption. Most systems are provided with a set of default values and use the host computer's operating system as a baseline on which to build a more extensive, and customized, EDM environment. Administration functions also include many other housekeeping activities, such as establishing and maintaining system backups and data archives. Such information, once defined, should normally be used by the EDM system to provide on-line information to the end-users. For example, if a user requests access to a particular data file (via the systems meta-database) which has been archived onto a tape volume, the system should be capable of informing the user accordingly and displaying the tape volume number to recover the file. The actual task of recovering the data file may rest with the administrator, the system operator or the end user depending on the size of the installation, the resources available and the operational procedures agreed. An EDM administrator's screen to define and maintain system users is shown in Fig. 8.17 by way of example.

EDM system architecture

We have previously considered the various types of user interface offered by different EDM systems, the users involved and the associated functionality, the input requirements and outputs screens, information and reports. But what does the EDM system look like internally? How is it structured and how do the individual elements interact with one another? Once again there is not a lot of commonality between systems. There are three main reasons why different EDM system architectures vary to such an extent: the scope of the system itself, the functionality provided, and the focus of the vendor who supplies the solution. Each individual offering on the market, although generally based on an RDBMS, will usually differ in the way the generic functionality is provided, for example security and control functions, tailoring and customization, process control and management functions. The two system architectures shown in Fig. 8.18 should therefore be considered as examples only—they simply serve to show how two individual system suppliers have decided to structure their internal architectures. It does not mean that others will necessarily be similar—each system offering will have its own advantages and disadvantages, just like the user functionality aspects we have previously discussed.

However, we can take heart from the fact that there are certain aspects which are common to many of the different solutions available on the market today. For example, all systems provide user interfaces which should be intuitive and easy to use. All systems have a central filing or electronic vault concept to store managed product data. Other main elements include a meta-database to maintain pointers, relationships and rules over the relevant data, application code to perform EDM-based functions such as initiating

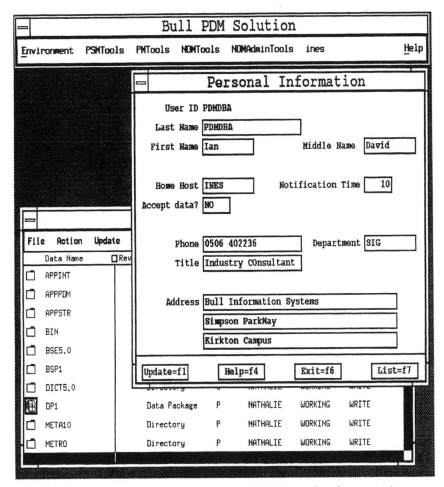

Fig. 8.17   EDM system administration—user definition and maintenance (*courtesy of Bull Information Systems*)

applications, checking rules and transferring and translating data, and communication and interface capabilities to other applications and databases.

System 'client' functionality is often delivered on UNIX and VMS, and on PCs such as IBM compatibles and Apple Macintoshes via server terminal emulation. Almost every EDM system is capable of running under the UNIX operating system for both client and server functions. Many support Digital's VAX/VMS for server functions, but only a few systems support IBM mainframes as servers.

Most currently available EDM systems require Ethernet and TCP/IP for successful network communication. VMS-based systems frequently support

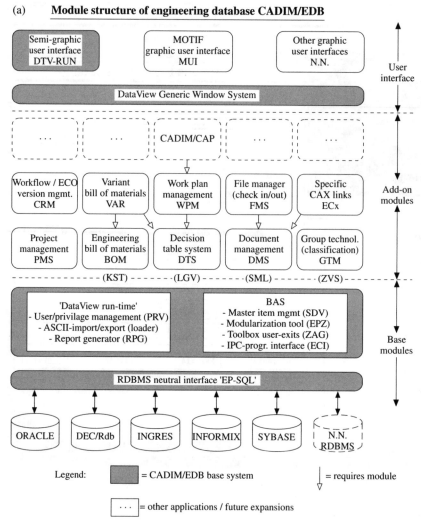

(a)   **Module structure of engineering database CADIM/EDB**

Fig. 8.18(a)   EDM system architecture examples (*courtesy of Eigner & Partner GmbH*)

DECnet. The example EDM architectures shown in Fig. 8.18 use 'network interface layers' to communicate between the individual system modules and the meta-database itself. Some vendors' systems are tightly coupled to specific database systems and can only use the network facilities of that system, other vendors have RDBMS-independent interface layers to enable them to support multiple database systems. This concept also extends to the meta-database itself which, as we discussed in Chapter 4, is normally based on an RDBMS which supports SQL (Structured Query Language). The most commonly used

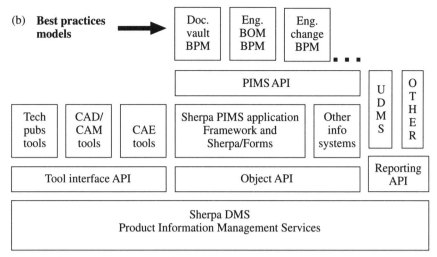

Fig. 8.18(b)   EDM system architecture examples (*courtesy of Sherpa Corporation*)

RDBMS systems for EDM applications include Oracle, Ingres, Informix and Sybase. Others include Rdb, Progress and a number of more 'proprietary' database systems. The majority of system vendors will offer their solution on a range of RDBMS systems, usually a common system for the entire implementation, but a few allow multiple types of databases to be intermixed and operate as if they were one.

Many EDM suppliers are now implementing object-oriented architectures and techniques where each different type of document/file can be modelled as an 'object'. Each individual object belongs to an object class and can be defined with different characteristics which determine its behaviour under certain conditions. For example, a 'plot' option may work for a graphics file and be accepted as a valid operation, whereas it would not make sense for a directory.

Finally brief mention should be made, before we leave this section on system architectures, of application program interfaces (APIs). APIs are used in many areas of computing technology but in relation to EDM are the mechanism whereby the applications themselves (e.g. CAD or DTP) can be closely integrated to the EDM system itself. In this way, users can operate as normal in their preferred application and, via the normal applications user interface, can access EDM system functionality as if they were using it directly. This is mainly used for central filing operations via the electronic vault, database enquiries and product structure details. The use of APIs should not be confused with the concept of encapsulation which we discussed earlier. Encapsulation allows an application to be launched from within the EDM environment and user interaction tracked accordingly. It is much easier to achieve and does not require the application to be modified in any way. It therefore allows a user to 'integrate' an application into an EDM environment over which he has no direct control (or source code). APIs on the other hand operate at a much lower level and do require access to certain application code subroutines if they are to be 'integrated' to the required degree. They do, however, provide a very tight interface when used correctly, and are used by the suppliers themselves when integrating their own software into the EDM system, as well as being made generally available via subroutine libraries. In this way 'seamless' integration can be achieved where the user is unaware that he is moving from one application to another (e.g. from an FEA system to the EDM filing routines) and commonality of system interface is preserved.

8.3 The pilot system

We introduced the concept of the pilot implementation technique in Section 7.5 together with the two other main implementation approaches, the 'parallel' approach and the 'push button' approach (often called the 'cold turkey' method in MRPII-speak). This section will take these introductory comments and expand them. For example, what exactly is a pilot and what does it consist of?

In strategic terms a pilot can be considered as having three main elements, each of which requires to be addressed if EDM is being considered as part of a change management/BPR exercise:

1 *An organization pilot*   Consideration of any changes which require to be made to the organization to accommodate EDM, for example the creation of a Chief Information Officer (CIO) to provide the required control and direction of the company's information.
2 *A process pilot*   Reviewing and changing any necessary processes identified as being required to streamline the EDM process, for example changing from a serial to a parallel review and authorization process.
3 *A system pilot*   A method of introducing the actual software solution itself in the most structured and seamless manner possible. Basically, this involves selecting a group of products, or one, or indeed one part of a large product (as long as it contains a reasonable number of parts) and operating the proposed new system on it, while controlling the other 99 per cent of the company's products in the normal manner. It reduces the risks associated with the introduction of any new technology on live product design and manufacturing data by proving that the system works before fully committing the company to it.

An additional advantage of this approach is that it facilitates and encourages everyone concerned to understand the product or technology concerned and learn about its use in a real-world, yet controlled, situation.

Pilot approaches encourage the systems concerned, in this case EDM, to 'work' from the earliest possible stages.

Traditionally, many IT and DP personnel have often disliked this approach. This was probably due to their perception that the majority of systems effort (i.e. installing and 'tweaking' it) had already been done leading up to the pilot, and the system should now be used live. While this is true, the pilot approach allows a less dramatic introduction and is much less likely to cause major operational problems should a problem be found in the initial live runs.

There are generally three stages involved in running a system pilot approach:

- *The computer pilot* This involves proving the computer configuration upon which the solution is based, i.e. ensuring all relevant aspects (hardware, software, networking and communication facilities, databases, etc.) are up to the job ultimately expected of them. It should also involve checking software functionality, perhaps against an agreed testing/acceptance schedule, loading test data for trial runs and timing tests, and so on.
- *A learning and consolidation period* This period normally runs after the computer pilot has been successfully completed, since it relies on an installed, available and robust computer resource. However, depending on the situation and the wishes of the company concerned, certain aspects of it can be run in parallel, perhaps utilizing a different computer platform as an interim step. This phase concentrates mainly on education and training and consists of 'walkthrough'-type exercises aimed at increasing familiarity and planning for future possible scenarios. It is often termed a 'conference room pilot', and although not always held in a conference room, does serve to indicate the off-line nature of the work undertaken in this phase. All those concerned with the introduction of the system should be involved at this stage: suppliers, internal project team members, systems analysts and consultants. The aim is to model various business scenarios and prove the operation of the system under real-life situations. System functionality should be agreed and, if necessary, modified prior to the full pilot run.

   This should help establish both the operating and system procedures and identify areas that may require policy direction changes. This is not product training (although that will generally be required as a prerequisite), it is 'simulation' training: getting all groups round a table regularly and working through a hypothetical situation to ensure that the system can cope.
- *The 'live' pilot run* The time leading up to, and during, the live pilot run is usually the most tense—tempers become frayed and nails bitten. This is the stage when the system is run live right through the selected trial data, the objective of course being to prove that the system works to a satisfactory level to enable an eventual complete 'switch-over'. Although this phase is

very important, the majority of problems should have been envisaged during the conference room pilot stage and appropriate measures taken to address them. However, no matter how good the planning is something generally goes wrong and it is important that everyone concerned, as well as expecting the unexpected, is trained and prepared for this period, in order to minimize any disruption to the operation of the company concerned.

When considering pilots, the question usually arises, 'What should the live pilot run consist of?' The answer of course depends on the nature of the system being introduced. From an overall perspective it is important to tackle the major functional areas as early as possible. This helps to:

- Focus top management's attention on the situation as a major strategic issue
- Attack the areas where most benefits will accrue, thus easing the justification of the overall project
- Provide visibility for the overall project and encourage EDM to be used by all company departments

The pilot itself can also be 'scalable' and related to each specific stage within the overall plan (as we discussed in Section 7.5). For example, there may be a pilot to trial the initial use of the electronic vault, a further pilot to establish workflows and perhaps a final pilot to extend the EDM solution into the areas of product structure management and CM.

There are also a number of more general points which apply in most situations. For example, the following points need to be considered:

- *Dataset size*   Choose a product (or subset) with a representative number of items to ensure meaningful test results. In the case of EDM systems this means choosing a suitable product structure (i.e. one with enough sub-assemblies and parts) and associated data files. It is usually best to limit the risk here by starting in a fairly small way and growing in line with the increasing knowledge and experience gained.
- *Product*   Where possible choose a product which represents the majority of item types you normally deal with, but be careful that the pilot does not grow too large as a result.
- *Cross-section*   Try to select as wide a ranging group of items as possible with the chosen product. Once again a balance is required here, because there is often a tendency to bring too many options into the pilot concerned. If there are many groups of items to be considered, it is often more advisable to separate them out into a number of different pilots rather than attempt to consider them all at the one time.
- *Self-contained*   Try to select a product which is as self-contained as possible to reduce the need to expand into other product structures to give meaningful results.

The pilots and their associated data should also attempt to relate in some way to the original objectives set for the system (refer to Section 7.3). This is not always possible and some objectives will only be met after much of the overall system has been installed and operational for some time, but tying-in any objectives to an early pilot will provide initial visibility to management that the objectives can be and are being met and, when achieved, will provide a much-needed 'morale booster' in the initial stages of the project.

If the pilot run is successful, plans can be made to transfer the live operations to the new system. As we have stated, though, problems will generally be found and it is important to pursue these vigorously. The necessary changes should be made and the pilot re-run. The 'live' pilot should be run for as long as it takes and the project leader should ensure that everyone understands that live use of the complete system will not go ahead until the pilot runs to everyone's satisfaction. In this way no-one should feel 'railroaded' into accepting a less than optimum solution.

When moving all operations onto the new system (often termed 'cutover'), it is best to perform a series of multiple 'live' runs each getting sequentially larger until all items are operating (although all data requires to be initially loaded) rather than dumping all the requirements on and doing a 'push button' job. During the entire project, but especially when moving all operations over for live use, it is vitally important to *communicate* what is happening to all team members, users and the steering committee. It is important to keep any consultants, and the system supplier, involved and informed at all times. Good communication ensures that costly mistakes are not made when things go wrong and establishes a feeling of achievement when things go right.

In general, then, successful EDM implementations, like many other IT-related projects, utilize either a pilot approach, a modular approach, or a combination of both. The modular approach (sometimes also termed a 'staged' implementation) was discussed in Section 7.5 and relates to the use of individual software modules, i.e. even the initial pilot may not include all software modules purchased—some may be brought on-line at some later stage as the project develops.

8.4   Advancing the use of EDM

It is important that any business enterprise considering implementing an EDM solution clearly understands its motivations in doing so. While the generic reasons are now well understood, where everyone is trying to create and deliver better-designed, higher quality products in less time and at lower cost, each organization will also have more specific reasons for investing in EDM. These might be, for example, to assist in conforming to internal operating procedures, reducing scrap, rework and redundant part inventories, and minimizing the overall number and associated cost of engineering changes.

Many of these motivating factors are also interrelated and cannot be considered in pure isolation. It is therefore important to understand and prioritize all motivations that drive a specific EDM implementation.

It must also be remembered that technologies such as EDM should not only address these types of requirements, which are either current or in the near future, but should hopefully be flexible enough to assist in longer term aims.

Therefore EDM should be thought of as a dynamic technology which can be continually added to as related technical advances are made and adapted for commercial use. For example, in many companies improving productivity and flexibility, while reducing the associated training effort, is an important factor. EDM systems relieve users from having to learn detailed operating system and communication commands. Users are guided through their part of the product life-cycle and can initiate the appropriate tools across systems and applications without having to re-learn new environments. This may then lead onto the related benefit of improving a user's ability to find, access and share data and information in a distributed environment, which in turn develops into the key area of improving the levels of integration between engineering and manufacture and assists in the formation of a concurrent engineering environment. Ongoing refinement and development of the system may then lead to it being a supportive element in the company's drive towards ISO 9000 and perhaps further onto TQM.

By considering EDM in this flexible, dynamic and continually changing manner, we can see the strength of support it can provide at the highest strategic level. EDM means many different things to different people. For a small organization the user interface, ease of implementation and standard application interfaces may be of most interest. For larger company-wide implementations, the ability to tailor the system, the provision of APIs, effective administration facilities and distributed, heterogeneous support may be more important. Defence-related companies may be concerned with CALS and other standards, a multinational company with multilingual facilities and so on.

But EDM technology can also provide elements of support at other lower levels, such as departmental and user level. For example the IT department of a company implementing EDM may employ some highly individual and creative employees who are used to having their own development environments and tools which may not communicate with those of their colleagues and conform to company standards. The EDM systems development environment may enable such individuality to be maintained but also provide a structured environment to exist via the use of a standard RDBMS, 4GLs, APIs, standard libraries, etc. perhaps leading to the structured use of OODBMS technology. Similarly, individuals within an engineering department may rank the user interface aspects very highly and advancing EDM use within such an environment may involve taking this improved UIF and using it to provide easier-to-understand product structure representations. These may be used within ECRs. The improved communication and reduction in misunderstood ECRs may raise the user's, and department's, profile in the company's more strategic goal of implementing ISO 9000.

Finally, in the general advancement of EDM system use, there are many reasons to be cheerful. The systems currently available on the market provide comprehensive functionality and allow a wide range of interesting related technologies to be brought into play. The project management of an EDM system should therefore be anything but dull! It can provide a very interesting and stimulating environment for all those concerned, from the highest to the lowest levels in a company. In addition to the technical aspects, there are many other tasks which require to be considered. For example, maintaining the enthusiasm of the staff and project team, ensuring the momentum of the project is not lost, developing any enhancements in a timely and effective manner and so on, are all tasks which should not be underestimated.

However, we can minimize the risks involved by learning from those who have already implemented EDM, and from the general experience gained in related IT projects. This aspect of communication is key. Anyone considering implementing EDM with little or no prior knowledge of the current marketplace should familiarize themselves before committing significant resources. There are many specialist awareness seminars, roadshows and exhibitions on the subject which can be used to provide a suitable basic knowledge. This can then be supplemented by more detailed information gained from consultants, suppliers and market analysts.

As with any rapidly developing technology it is important that the information on which you act is both relevant and up to date. Consultants will generally be aware of the latest market offerings; this is important because of the very rapid product development cycles encountered. This is of particular relevance when evaluating products—you must ensure that you evaluate the current version of a product. Just because a product was evaluated a few months ago does not mean that you have a valid understanding of its capabilities today. The same situation applies when talking to existing user sites who may not be using the most up-to-date version of the product concerned.

8.5   Interfacing and integrating with EDM

Before we leave this chapter on the more practical aspects of EDM, i.e. what it looks like and how it is used, we should discuss one further important point: the interface and integration aspects of EDM to and from other software applications.

The question here is whether this subject should be included in Section 8.2 as part of the currently available system features, or in the next chapter on future EDM developments. The reason for the dilemma here centres mainly on the subject of EDM/MRPII system integration. In many respects this can be considered as something which, although heavily tailored, can be provided today. On the other hand, it is a subject of major importance to many prospective and existing users of EDM who require a simpler, more structured, and 'standardized' interface mechanism or toolkit to reduce the currently lengthy timescales and correspondingly high costs of these tailored interface mechanisms.

EDM system suppliers will therefore require to commit considerable resources to its standardization and development. It is for this reason that it is considered as a separate section of this chapter.

Firstly, we should consider the difference between integrating and interfacing. Interfacing between two systems is usually taken to mean a loose connection, mainly by file transfer, and perhaps done on a batch or timed basis. For example we may have a sales order processing (SOP) system which operates on machine A and accepts orders from field sales representatives on a daily basis. These may then need to be fed into a production system on machine B for the daily production run the next day. In this example, an appropriately formatted interface file may suffice. This could be downloaded automatically at, say, 22.00 hours every evening onto the production system.

Integrating two systems, on the other hand, is usually taken to mean a more detailed form of connection between two or more systems, normally at a lower level in the database or program code. For example, looking further at the previous example, if our SOP and production systems were truly integrated, our sales engineers could perhaps link through from a sales order entry option directly into the stock module to check availability of the item(s) concerned prior to accepting the order. Perhaps stock could also be allocated to the order and delivery dates given to the customer at the time. Such integration can normally best be achieved if the two systems concerned are written in conjunction with each other and share the same coding, file and database structures. But it is also possible to obtain a similar level of integration via database, file and message handling mechanisms and is a task made much easier by the use of standard APIs, as we discussed in Section 8.2.

In general, integration of EDM systems with additional applications can be considered as being on one of three levels (note that the data which is being transferred is the codified data, i.e. the metadata of the EDM system and the structured or factual data of MRPII, and not the documentary data which only the EDM system can refer to and manage):

- *Application integration using APIs* Such integration tools can be used to create very tight integration between an application and an EDM system, where users can work entirely within their chosen application and, via the application's own user interface, access the EDM system. Figure 8.19 shows an example of a graphics application, a user's 3D CAD display, overlaid by EDM browser and information windows.

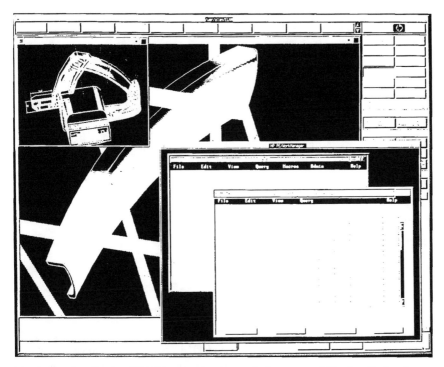

Fig. 8.19 Application/EDM integration example (*courtesy of Hewlett Packard*)

Certain other aspects of close integration often required between many CAD and EDM systems, such as those of drawing information and parts list (or BOM) transfer for example, can also be provided by the use of APIs. All drawings will contain basic information which may be held locally or in the EDM systems meta-database. Also, many drawings of assemblies or sub-assemblies will contain a listing of the parts used and 'bubble' identification numbers to relate them to the actual graphical representations themselves. Such details, as shown in Fig. 8.20, should be able to be optionally loaded automatically to and from the EDM database when the drawing is called up on the CAD system. Suitable mechanisms will of course be required to cater for update and change information. The exact operation of facilities such as these will depend entirely on the CAD and EDM systems concerned—usually the operation of an EDM solution sourced from a CAD vendor will provide the highest level of integration in these respects.

- *Application encapsulation*   As we discussed in Sections 8.1 and 8.2, encapsulation provides a moderate level of application interface by session tracking and logging mechanisms. Such integration is easy to achieve and does not require modification of the applications themselves.
- *Document registration*   This simply refers to the method of registering

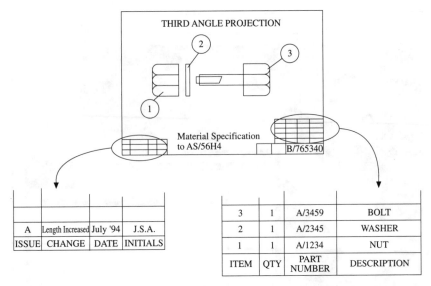

Fig. 8.20 Assembly drawing information

documents that have been created through external methods and are simply referenced by EDM. For example, the EDM system can manage files which are not interfaced to the EDM system in any way, as well as physical, i.e. non-digital (perhaps paper-based), documents which may simply be registered for filing via EDM under a drawer or filing cabinet reference.

The majority of current EDM systems linked to MRPII are interfaced rather than truly integrated. They generally operate by the regular file transfer of data from one system to the other. Although this sounds fairly basic, the majority of information can be handled in this way and the operation is usually automated and hidden from the user. The information currently transferred by such means is usually of two main forms:

1. The generation and modification of part/item records, i.e. the part or item data held which defines its existence, for example the item master record in the MRPII system and the corresponding definition in the EDM's metadatabase.
2. The definition and modification of structure information, i.e. the relationships between the various parts, assemblies and sub-assemblies, for example the BOM in the MRPII system and the product structure in the EDM system.

The problem in the transfer of such data, if it can be described as such, relates to data duplication. Both systems need to hold the same information to enable their successful operation. Short of making them operate together as one, on a single unified database, which one should be treated as the 'master'? Although MRPII vendors are currently moving into EDM technology (refer to Section 7.2) and no doubt fully integrated EDM/MRPII systems will be commonplace in the future (very few exist at present), they are currently perceived as serving two distinct markets. They therefore duplicate large amounts of data. For example, consider a design engineer involved in the generation of a new design of product. There will be a need to interrogate existing data to establish where items can be re-used, define new assemblies and parts where necessary and define the structures and BOMs required. All such information will need to be passed to the MRPII system to ultimately manufacture the product concerned, yet also remain in the EDM system for document filing, interrogation and configuration management purposes. When updates are made how will such information be transferred to the other system and how often should this be done? How will authorizations and concessions be handled in the MRPII system? There are three main options to the current dilemma of EDM/MRPII integration. These are shown diagrammatically in Fig. 8.21.

In option 1 the MRP system is treated as the master. The main role of the EDM system in this situation is to 'feed' the MRPII system and to use it for the majority of general enquiries, configuration records, structure details and part information enquiries required for everyday use. This option is currently the most common, especially in more product-centred environments, with a

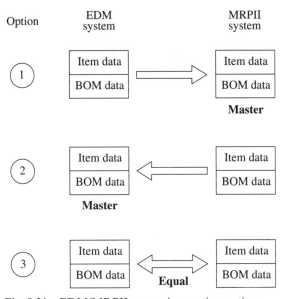

Fig. 8.21 EDM/MRPII system integration options

(by and large) one-way flow of data, with suitable validation and error handling facilities built in. A backwards link from MRPII to EDM may also be required but this is, in general, a more complex and error-prone operation. The KISS philosophy is best adopted here (Keep It Simple and it workS).

Option 2 is similar but treats the EDM database as being the master. Again a one-way data flow can be established and the procedures around the system can be configured to aim the majority of enquiries at the EDM database. In this case, however, the EDM system will not be able to handle the manufacturing-related enquiries, e.g. about the control of works orders and work-in-progress. These enquiries will have to be aimed at the MRPII system. It is for this reason, i.e. the uncertainty of where to look for specific information, that option 1 is generally favoured. However, this approach is more likely to benefit a more project-related environment where a basic configuration structure initially exists which is enhanced as the project nears completion, for example the building of an oil rig or a power station.

Option 3 treats both databases as being equal. While this is obviously the best solution overall (i.e. for both product- and project-centred environments), it does rely on both databases containing exactly the same information at the same time. Since they are only linked they would require on-line updates between the two—real-time interfacing, i.e. integrated! For example, perhaps, prior to placing an order for a spare part, a customer wishes to enquire about the appropriate version of the part in question. The person dealing with the enquiry may determine that the current version is version C, by accessing the MRPII system, and may be unaware that an urgent change has been authorized on the EDM system to update the item to version D because of a safety-related aspect. The MRPII database may only be updated once the next interface file is passed from the EDM system. Many companies would consider such a mechanism, even if performed every 15 minutes or so, to be totally unacceptable. Others may be able to accept it—it depends on the company concerned, their chosen marketplace and methods of working. However, to operate successfully in such a manner the organization concerned needs to understand the limitations involved, to be prepared to accept them, and to establish appropriate business procedures to handle them.

The problems of interfacing EDM and MRPII systems, complex and difficult as they are at present, will only become more so unless some form of standard interface library or toolkit is established (API) to enable their speedy and efficient generation. Market pressure will undoubtedly contribute here, as many more organizations will want to integrate without having to resort to expensive and cumbersome bespoke solutions. So too will the number and sophistication of the interfaces themselves. The present concentration on item master and BOM data hides the fact that there are many more areas where competitive advantage can be gained from a comprehensive interface, for example a strong relationship between customer records and configuration management records to easily show what customer has what build of product (this can be further integrated into a customer service system), or a combined costing module to enable material, labour and on-cost to be moved between systems where necessary to avoid data duplication and re-keying. Many MRPII vendors either already have, or are currently developing, additional modules to extend the basic concepts of 'traditional' MRPII into ERP and COMMS, such as Minxware from Spectrum Associates Inc. and Triton from Baan International. Detailed integration between EDM and MRPII technologies should therefore become a logical and natural progression as these systems continue to develop over the coming months.

9

The EDM marketplace and future aspects

9.1 The EDM marketplace

EDM has had a rather hesitant entry into the world of computer-integrated manufacturing. Many people discounted it as being an 'icing on the cake' technology and never really believed it would establish itself as an identifiable solution in its own right. In a way, this can be blamed on the technology itself and the fragmented way it has presented itself and grown over the recent years. But time has proved its value to many different kinds of organization and a steady, strong acceptance of the technology is now apparent. This has now grown to the point where it is considered to be a fundamental aspect of any advanced engineering environment, as we have seen. But how did this come about and how does the current situation compare to the market predictions previously made? To answer these kinds of questions, this chapter will focus on the EDM system marketplace and consider its future growth potential.

While other more traditional and visible aspects of engineering technology seem to have new market studies presented every other week, EDM has only had one formal European analysis undertaken to date. In 1989 Frost and Sullivan produced a detailed analysis of the European EDM marketplace in terms of business requirements, opportunities and available solutions. It proved invaluable at the time, for prospective users and suppliers alike, and has remained a useful and informative document since its release. Although the market has matured significantly since this time and the information in the report has been reproduced in various editorials and trade publications, no other study of a similar nature has since been undertaken. Smaller European studies and estimates of the size of the EDM market are available but are not comprehensive and are generally based on subsets of the overall market potential.

However, ongoing and up-to-date EDM market information is available on a worldwide basis. CIMdata Inc. are a US-based consultancy which specializes in EDM technology. In addition to EDM, CAD, CAM and engineering-related consultancy services, CIMdata produce a wide range of technology guides and reports (see Appendix A for full details). These include EDM pricing analysis and market service reports. By using both of these sources of information, we can build a revealing picture of the growth of EDM, its current status and its future potential.

Frost and Sullivan took a brave step in 1989 in predicting what could become a very sizeable market almost before it had even begun. The report itself not only contained the market analysis itself for the period 1989–1993, but also contained an introduction to EDM and its related technologies, details and types of EDM systems, benefits, implementation guidelines and profiles of EDM system suppliers. In 1989 the market for identifiable EDM technology was just beginning to appear. Until this time the majority of companies who were either considering, or had implemented, EDM were also mainly to the fore in the use of CAD/CAM and other engineering systems (e.g. automotive, aerospace, electronics) and the associated EDM solutions had a heavy bias towards in-house development. This was due to the fact that companies such as these were able to recognize the problem, felt they could not wait for the market to develop a more 'packaged solution', and were able and willing to commit their own resources to develop their own solutions. However, the need for EDM system technology had been known for some considerable time and the report suggested that, at least initially, the majority of computer system vendors were slow to recognize and respond to the need. During the period from 1987 to 1989, a number of packaged systems were launched, with the prediction that there would be many more. This was considered to be a critical time for the EDM marketplace—it was too early for many of the systems to have made many sales or gained significant market share, yet the revenues from such sales were badly needed to fund the often significant development programmes required to gain competitive edge. The report paralleled the situation at that time to two earlier technology marketplaces: the emerging CAD marketplace from 1977 to 1987 and the engineering workstation marketplace from 1987 to 1992. There were, however, some notable differences:

1 The CAD market was more focused—EDM was likely to appeal to a much broader target market.
2 The opportunity to sell support and services (a growing trend) was considered much higher with EDM than the then norm of 23 per cent with CAD.
3 To a certain extent CAD or CAM systems, because of their highly 'visible' nature, were considered to be much easier to sell than EDM, where longer selling cycles to a much wider audience were expected.
4 The workstation market was of course hardware orientated, with system software and services representing only 13 per cent and 18 per cent respectively in 1989.

Certain other aspects of EDM system technology were also noted, for example the drain or 'hardware drag' of larger database management systems which require ever-increasing amounts of processing power and disk capacity, the very large difference in system prices and how they are calculated and the bias towards US-based EDM solutions. Finally, and most importantly, the report highlighted the need for adequate system customization, consultancy, support and training. As we have seen, these are important elements if an EDM system implementation is to succeed. They not only need to be supplied by the various system vendors to a high standard, but must also be budgeted for by the customer. The report suggested that sales by system vendors, excluding any in-house development activity, was unlikely to have exceeded about $30m in Western Europe during 1989. It predicted that system take-up would be slow during 1990, although considerable activity in the form of feasibility studies could be expected. During 1991 the number of pilot systems was expected to grow rapidly and in later years large production systems would become 'numerous'. The total EDM marketplace was forecast as being worth $1350m by 1993. This trend is shown in Fig. 9.1.

This total was made up as follows:

Hardware	33 per cent
Software	25 per cent
Services	42 per cent

and the expected market breakdown by national area was as shown in Fig. 9.2.

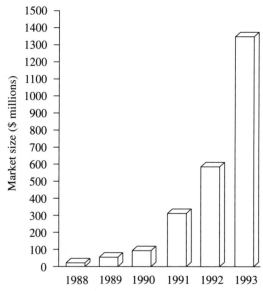

Fig. 9.1   The European market for EDM systems 1988–93 (*source: Frost and Sullivan, 1989*)

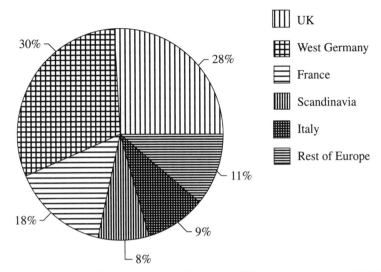

UK

West Germany

France

Scandinavia

Italy

Rest of Europe

Fig. 9.2   The EDM market by national area 1992 (*source: Frost and Sullivan, 1989*)

Although the overall report was of a very positive nature, certain key problem areas were identified, with certain aspects of one still apparent:

1   *Lack of EDM awareness and experience*   This was felt to be a very important bottleneck and manifested itself in two main areas:
    (a) The acute shortage of people with in-depth EDM understanding and experience. This has largely been addressed by many consultancy specialists and system suppliers.
    (b) The need for engineers and EDM decision-makers to become fully aware of EDM and its associated potential. This will ensure confidence in the technology and in its ability to provide a good return on the investment made.

2   *Robust and high-availability systems*   Engineering data is a very valuable asset and the 'up-time' of the systems which supply it (both hardware and software) is of paramount importance. The recent, and ongoing, advances in all aspects of associated EDM computer technology have helped to address this issue although it should still be an important consideration in any EDM purchase. Certain users have also given consideration to fault-tolerant computers for hosting EDM applications.

3   *Database system technology*   The comments made in point 2 also apply to the DBMS used within the EDM solution. The initial concerns here related to the user-unfriendliness and slow performance of certain database solutions available at the time. Once again these points have largely been addressed, not only by direct database technology and object orientated

techniques, but by the dramatic changes in hardware price/performance over the recent years.

4 *Cheap, user-friendly systems*   Once again the situation has improved here with windows technology and the provision of suitable customization and integration toolkits.

In many ways the current EDM marketplace and its future expected growth form a natural extension to the picture presented in the Frost and Sullivan report. There are no nasty U-turns or surprises to be found; this reinforces the stability of the market and its gradual acceptance and growth. CIMdata Inc. has also identified a relatively early marketplace which is beginning to grow and evolve. Its current findings indicate the following market trends:

- The emergence of EDM as an accepted international technology tool
- There are a growing number of large systems in use worldwide (some having in excess of 2000 users), thus confirming the benefits that can be achieved
- Many companies are now considering EDM and many more are in the early phases of evaluation
- CALS and ISO 9000 are major driving forces
- Major systems integrators are now becoming involved
- A strong focus on integration and workstation/server systems
- A growing number of users per system and systems per company
- Increasing use of standard/packaged EDM products
- An increasing systems expansion and add-on potential

CIMdata estimates that enterprises worldwide invested over $250m in EDM-related software and services in 1992. This amount does not include investments in related technologies such as imaging systems, or investments in hardware and networks required to support EDM. Current forecasts indicate a compounded growth rate of approximately 40 per cent for EDM software and service sales over the next five years.

Figure 9.3 shows the worldwide growth rates for purchased EDM software and services from 1989 to 1996.

Figure 9.4 shows the worldwide figures split geographically. By comparing the European totals with the original Frost and Sullivan market study, we can see that although the trends are similar, some interesting differences exist. However, before drawing too many conclusions from this comparison, it should be remembered that the Frost and Sullivan market analysis included hardware, software and services revenues whereas the CIMdata study concentrates on software and services only, and that the well-known saying that 'there are lies, damned lies, and statistics' also applies here in a generic sense.

First, it would seem that the initial market totals in the Frost and Sullivan study were slightly on the high side, even allowing for the hardware element.

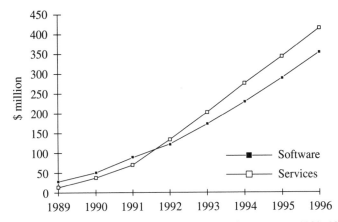

Fig. 9.3   Worldwide EDM software and service revenues 1989–1996 (*source: CIMdata Inc.*)

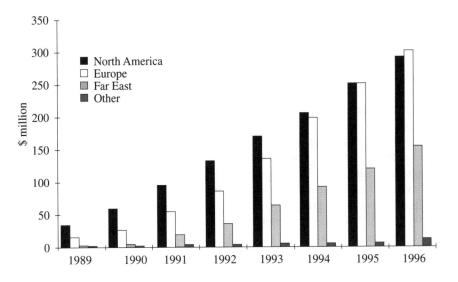

Fig. 9.4 Worldwide software revenues by geography 1989–1996 (*source: CIMdata Inc.*)

Secondly, the growth rates would seem to have been significantly overestimated. This should not be taken as a slur on the Frost and Sullivan report—any attempt at guessing futures, especially in a fast-moving technological environment, is bound to prove difficult. The differences in 1993 revenue comparisons, which are in excess of $1000m, can be partially explained by the then difficult to quantify, and rapidly changing, hardware/software/services mix and the fact that the study was conducted prior to the worldwide economic recession of the early 1990s. The fact that elements of the technologies excluded from the CIMdata study may actually be included in the original market study could also play a significant part in accounting for the difference.

CIMdata's market information also highlights some other interesting facts about the EDM marketplace. For example, although a tremendous amount of flexibility is required to adapt EDM systems to the various operating environments which are encountered, there is still a great deal of commonality between the capabilities needed from one organization to another. As a result, prepackaged EDM solutions and their underlying architectures are becoming more consistent.

The EDM marketplace would seem to be exhibiting similar characteristics to many of the leading-edge technology solutions of the past, where the USA initially led the field—for example CAD/CAM and MRP/MRPII. However, just as with the eventual growth and take-up of these technologies in other countries, EDM systems are now being developed and released by many non-US suppliers, primarily within Europe. Also, although EDM activity and the number of implementations is still larger in the USA, the European market is growing at a much faster rate, with Germany in particular taking a significant share in this overall market.

Finally, the use of EDM technology has moved from being centred on aerospace and electronics (the original main proponents) to other industrial sectors such as automotive and general manufacturing industry. But EDM can be applied to a wide range of industries, not just manufacturing. It has been successfully utilized in the electricity supply and petrochemical industries, and is now beginning to migrate into other non-discrete manufacturing areas, such as pharmaceuticals. CIMdata expects this trend to continue as the technology is developed and adapted to the special needs of different industrial sectors.

In addition to general EDM market information, CIMdata Inc. also provides detailed market information, specific evaluations and comparisons of commercially available products and a full PDM (Product Data Management) Buyers' Guide. Contact names and addresses are included in Appendix A.

9.2   The future of EDM

You may have read some introductory textbooks and felt that the subject in question was presented in some detail and a reasonable understanding was gained, only to find out later that what was actually presented was only the tip of a very large iceberg. EDM technology will probably prove to be similar in many respects. Although the basic concepts, procedures and solutions are broadly similar there is a wide variation in individual detail, how the various facilities are provided, and the related technologies are sometimes complex and wide-ranging.

This is one of the main reasons why EDM is such an interesting and addictive subject—there is always some new aspect or technique to be discussed and learned. Another reason is its dynamic nature—it is still very much a growing marketplace. Indeed, the analogy with the CAD market a number of years ago is very relevant. Many readers may remember the excitement of being involved in CAD when it was 'leading-edge' technology, growing in acceptance and use at an incredible rate. Environments such as these are always exciting and fun to be involved in.

Although the CAD market has now stabilized and matured, the EDM marketplace is exhibiting the same appeal and dynamism as the CAD market did some eight to ten years ago. It is important, however, in looking at the growth of EDM and drawing analogies with other markets, that the underlying trend is considered and not just who or what 'flavour' of EDM technology is ahead at any particular time. A leadership position is not permanent. If we were to consider a list of the EDM market leaders from, say, two years ago, it would contain a completely different group of products and suppliers from a current listing. During this time some new leaders have emerged, while other former leaders have lost considerable ground. Other generic and external factors also play their part, such as economic factors, the change in the world's centres of manufacturing, the demise of defence-related spending, and so on—all of which have an influence on engineering and manufacturing industry.

Having presented EDM in its more global context, introduced and considered the underlying technologies, and examined the detail of EDM systems at a more practical level, it is interesting to conclude by speculating on some possible future scenarios for EDM. It was stated earlier that any attempt to predict the future is a rather risky undertaking, but if we do not consider the future, how can we plan for it?

Before detailing a number of case studies in the next chapter we will therefore conclude our 'mainstream' look at EDM by considering some of the advances which could be expected both within the field of EDM itself and some of the wider enabling technologies and their relationship to the advancement of general computing technology.

With regard to EDM technology in particular there are a number of areas where we can already see trends beginning to emerge. For example, the use of graphical user interfaces (GUIs), an area of rapid growth and user acceptance over the past few years, is expected to continue. Early EDM systems, perhaps because of the display technology utilized, were unable to cope with the graphical representations of product structures in a fast and efficient manner. If such facilities were provided at all it was generally by a slow database interrogation and display on a CAD terminal. Today, such representations are considered standard functionality. The use of generally available technologies such as MS-Windows, X-Windows, OSF/Motif, etc. and standard user interface (UIF) libraries have meant that system suppliers can now provide such functionality in a fraction of the development time it used to take.

The advances in UIF technology are already showing movement in the area of multimedia. Cimlinc's Linkage product, for example, can be used to combine images, text, voice and video simultaneously from standard DBMS interrogation tools. EDM systems of the future may well be able to take advantage of such technology and, in addition to displaying the BOM, could perhaps ease the task of visualizing complex configuration management records or showing images of a parts change control history, or even a video of its manufacturing or assembly processes.

The subject of object oriented databases in relation to EDM is a further area undergoing rapid development. Although OODBMS technology is still very young, the indications are that its use could provide significant performance improvements in the often heavy processing requirements of traditional engineering data applications. Currently, all of today's major EDM systems use RDBMS systems to store their metadata. In general they cope well, but response times can slow down considerably when a large number of simultaneous user requests are made.

OODBMSs are designed to combine the functions found in RDBMSs with the capabilities of object-oriented programming environments, and offer the potential to improve the performance of many EDM functions. However, although their potential is very high, there is no history of how well they will actually perform under real load conditions. Some time will therefore be required to further develop and tune EDM systems based on OODBMS before their widespread use.

Movement of EDM out of pure CAD or engineering environments, together with the increasing interfacing and integration with other software, is another area of future growth. Many early EDM systems were developed by CAD vendors and the terminology and data models used reflected mechanical engineering cultures and working practices. Most disciplines, however, need to use a set of common data functions (e.g. creating, viewing, changing, approving) and as a result EDM technology is moving into other environments where the EDM (or PDM) acronym should probably give way to a more generic 'DM' where there is a more general need to marry the electronic and paper worlds in new ways to match business needs. For example, software development environments have for many years adopted the concepts of the electronic vault (source logging) and configuration management. We shall probably see an increase in such market penetration by EDM vendors over the coming few years—the current limitations are not generally product related but more related to the market focus of the vendors and their associated ability to support such new markets.

Currently, interfaces between EDM systems and other application software, such as analysis, financial, scheduling or manufacturing systems (MRPII), are as we have seen—at a somewhat basic level. However, the number and complexity of these is expected to rise over the coming months as EDM is further accepted and developed. Such interfaces would typically include:

- Mechanical CAD systems
- Electronic CAD and PCB design systems
- Engineering analysis (FEA) systems
- Document production systems
- Project management systems
- Stock control and MRPII systems
- Specialized applications
- Asset management and financial systems

and the development of an increasing number of standard interfaces to legacy systems and individual 'point' solutions.

A trend which has already been established and is expected to continue to grow is the number of large-scale EDM implementations. The database and technical advances over the coming months, in conjunction with the advances in EDM itself, will probably result in the emergence of some systems having several thousand users over an enterprise-wide installation. We have already seen the similarities of the current EDM market to the MRP market of some 15 to 20 years ago. The expectation of system size is in line with this analogy, as is the growing and continuing acceptance of prepackaged applications and a convergence of functionality.

Another aspect of EDM which is in the early stages of acceptance and where future growth is likely is EDM 'frameworks'. A framework is a common user environment for the cooperative use of a set of tools to support a specific type of user. Frameworks can be used to guide users through the correct set of steps, applications, data control, etc. via a common user interface. Although EDM systems were not initially designed as complete frameworks, it can be seen how the currently available EDM technology has grown to the point where they are now becoming 'enterprise' frameworks, and not just for ECAD or MCAD (electronic and mechanical CAD) subsets. This trend is expected to continue with EDM frameworks assuming more visibility and functionality, with the common overall environment allowing individual application vendors to concentrate on their own particular area of expertise.

The picture starts to blur when we move the timescales out and begin to consider the wider aspects of general computer technology and applications. Who would have thought only ten years ago that you would be able to get a pair of spectacles from a computer-controlled machine in less than one hour, that up-to-date technology such as word-processing and project management could be so easy to use, and that the PC on your desk would have up to ten times the power and a small fraction of the size of the then state-of-the-art minicomputer? If this rate of development is maintained where will technology lead us over the next ten years?

Most industry predictions would seem to agree on the continuing increase in computing power combined with a reduction in size. Apple have already developed a Personal Digital Assistant (PDA) which can receive and download information to a host system via wireless means. Peripheral equipment such as disks and communication hubs are already much larger than the processors they are attached to. Many also predict a growth in 'wireless' LANs and 'wireless' communications at all levels are expected to become the norm, fuelling the increase in multimedia's ability to present integrated voice, video and text. This in turn both supports and encourages the rise in home-based workgroup computing. Such communication and ease of use will mean that many tasks should be able to be accomplished on an individual basis, but with fast, accurate access to common shared databases, applications and people, via data and video links. Voice-controlled systems have been around for many years now and although they have performed well in niche applications, have gained little mass-market appeal. This is an area that analysts believe is poised for significant growth and general acceptance. If the concept of 'point and click' on mouse-based systems is replaced by 'look and speak', then the focus is centred more on the effectiveness of the people using the technology rather than simply the technology itself.

There is also the question of the paperless office, the boardless drawing office, etc. Will they ever happen? Mention the paperless office to the word-processing pioneers of the early 1970s and the response is usually an embarrassed cough followed by a long-winded explanation about an unforeseen enthusiasm for having everything written down. Certainly the environmental pressures on conserving resources and tree preservation will undoubtedly have their effect. When considered in conjunction with communication and other advances then the need for paper-based information will decrease significantly. However, the world is not yet ready for paper-free business and it is highly unlikely that manual and paper-based systems will ever be completely replaced.

Finally, there is the future of virtual reality to consider. Many of us probably only view virtual reality as a form of games technology, and although its initial use was aimed at this type of environment, the possibilities of its eventual development potential are almost endless. Examples are: videoconferencing anywhere in the world from your own office or home as if you were actually there with your colleagues; on an engineering front, being able to simulate and view the design or manufacture of a product before it is physically generated; 'virtual' training on an expensive item, viewing a pipework or electrical network and being able to 'fly' around it to investigate faults. By expanding our preconceptions of this technology, which is still in its infancy, we can see that it will probably become one of the leading technical developments of the coming decade.

Many of us may consider such future developments to be innovative and exciting, while others may dread the changes they will inevitably bring. Whatever your view, one thing is clear: changes, of whatever nature, will occur. We must endeavour to learn and use them to the best of our ability and to the benefit, rather than the detriment, of mankind and the environment.

10

Case studies

The original purpose in including a chapter on case studies was to provide detailed, practical information on how certain companies have approached and implemented their own EDM solutions—to act as a balance to what many may consider to be the more theoretical aspects discussed thus far.

The intention was to include three types of case study: the first a large-scale strategic EDM implementation to demonstrate its use at a multi-site, corporate level; the second to provide an insight into EDM use within a medium-sized engineering enterprise; and the third to describe a smaller file management or DO-centred system, to highlight the scalability of the concept of EDM and how, given a suitable longer-term growth path, it is possible to start small and grow a system in line with future needs.

However, the case studies which appear in this chapter do not follow this logic for a number of reasons. Firstly, in order to arrive at those EDM installations detailed here, a large number of companies were originally contacted. After speaking with a number of such companies it became clear that the systems could not be categorized into such predefined 'slots'. Each individual EDM solution, which took many forms, degrees of sophistication and levels of implementation, were all seen by the companies concerned as being of vital importance, and in the majority of cases formed a critical element of a much larger computerization strategy. The two case studies which follow are therefore presented on an individual basis and should be considered on their own merits and in relation to the company environment concerned.

One further main point regarding why there are only two case studies relates to the fact that of the many companies originally contacted, a large number consciously refused or subsequently found themselves unable to contribute. Some were able to provide overview material, sometimes generated by the EDM vendor concerned, of a sales-related nature. Detailed information was difficult to come by. This was a very interesting aspect which I have subsequently established and confirmed with other data management specialists as being a general problem.

It would appear that the majority of companies are unwilling to share their detailed experiences in such areas—perhaps they feel they have gained a serious competitive advantage which they wish to guard, or perhaps the more traditional entrenched and introspective engineering psyche has once again come to the fore. At least two companies who agreed to provide a case study subsequently changed their minds after the text had been generated, apparently because of the difficulties encountered in obtaining management and marketing approval—perhaps this is the major stumbling block! Whatever the reason, two case studies were eventually sourced and have been fully detailed over the rest of this chapter.

My deep gratitude must go not only to the companies concerned, but also to the individuals who helped with the material and authorized the use of the final text. I believe the effort in obtaining and detailing these two EDM system examples has been well justified. They both highlight and reinforce, in a practical and straightforward manner, the messages already presented throughout the text of this book. Since this was the main original aim of including them, I feel the effort involved to be worthwhile and their ultimate inclusion an undoubted success. I do hope they prove to be as interesting and informative for you to read as they have been for me to document.

Racal Avionics Limited

Introduction
Racal Avionics Limited develop and manufacture airborne computers, flight management equipment, satellite-based telephone systems and ground navaids at three sites in the United Kingdom, and employ around 500 people. In 1990 the company operated a manual drawing office with 20 staff. According to Mike Clay, Racal's Technical Services Manager: 'There was very little visibility and many of the changes were taking months to be processed through perhaps six departments. In total it was costing us a lot of wasted time, effort and money. Engineering change notes took months to process.'

Only electronic designs were created on a CAD system, and the problem was in managing the changes rather than creating the designs. Purchasing an EDM system was the obvious solution and this case study provides a comprehensive history of the development of Engineering Data Management within Racal Avionics Limited (RAVL).

Background information—the need for EDM
In the early 1970s a group component database, called COAST, was initiated, running on a mainframe. The only drawing office access was via volumes of hard copy, eventually replaced by microfiche. It was some years before RAVL acquired a single terminal. Data was manually copied onto item lists—a slow and error-prone activity.

The database was centrally controlled and suffered from a slow response to user-initiated system development requirements. Between fifty and a hundred sheets of new and modified item lists were being produced per week at that time and the process of researching data and manually entering it onto paper was time consuming and error-prone. Forty draughtsmen were employed, including contract workers, and it was necessary to run a checking section to ensure acceptable accuracy.

The problem became worse when a Racal Redac Mini-Maxi computer was installed to produce circuit boards faster (so that hand lay-up was largely abandoned). It was apparent at that early stage that, as data held on COAST was exhaustively checked at the point of entry, if item lists could be produced directly from that data, the accuracy of the item lists would not only increase at no extra cost, but, bearing in mind that there were around 200 components and frequently more on an item list, the time taken to produce the item list would also decrease. However, there were no commercially available programs on the mainframe which would achieve what was required, and the centralized Group Computer Services were unable to produce one.

In the late 1970s a stock control and purchasing system, called RIMS, was installed. Although a significant amount of data was entered, the system did not prove successful and was replaced by an ICL MRP system (OMAC) which was installed in the early 1980s. RAVL employed a 'database controller' who manually entered all the relevant data from the current item lists. At that time around 50 000 sheets of A4 item lists existed, and decisions had to be made as to what was going to be loaded and how this was to be achieved. Current production was already largely covered by RIMS and the transfer of RIMS data to OMAC was not as easy as anticipated and it needed considerable massaging before it was acceptable and accurate for OMAC. At this time the database controller commenced to make himself unpopular by refusing to corrupt the system to accommodate local problems produced by the so-called 'short cuts'. This incorruptibility produced dividends later, which provided an object lesson in maintaining the integrity of data.

Around 1988 there was pressure to reduce overheads at a time when the backlog of engineering changes was in excess of 2000 (and increasing), while the DO staffing level was around 18 draughtsmen and the database controller. Manually there was no visibility of the total quantity of changes as there was no central repository. Data searches had to be made against each piece of equipment drawn (which totalled a few hundred). The magnitude of the problem and thus the effort required to solve it was not known until an effort was made to enter all DCFs onto a system. Initially the data was entered onto DELTA (a PC-based database), with some difficulty, which exposed the problem but offered no solution. RAVL had continued to keep their objective in mind and had developed a Functional Requirement Specification (FRS). Based on this, and recognizing that the arrival of workstations and file servers offered greater flexibility under local control, they started to search for a system which would meet the FRS and which would have the capability to enable in-house development by relatively unskilled operators. A basic requirement was that all data must be entered only once but could be available in whatever form a user might require, bearing in mind that at one time each paper item list was being rewritten around the company five times in various forms with the consequent probability of errors. The company considered that any system chosen should be capable of interfacing with existing systems over a network, and that network should be capable of extension to cover all future areas of operation.

System selection and implementation
RAVL searched the marketplace, but at that time could not find any systems which met their requirements or could handle the volume of data required in a multi-user environment. A promising system based on ORACLE was ruled out as the vendor was considered too small to guarantee adequate support. One other emerging system of the time would not handle the quantity of data. Apollo and DEC had no system but offered their full support to RAVL to develop their own. However, the company felt it had inadequate knowledge, resources and time to start such a development from scratch.

ICL then offered DOMA. This system was originally developed by the CADCentre in Cambridge and was marketed by ICL. It went a long way towards meeting the stated requirements and ICL assured Racal they could not only develop it to meet the full requirement, but train RAVL to develop it further themselves. It was demonstrated with working interfaces to COAST and OMAC. By this time the Mini-Maxi had been replaced by Visula and the interface to take Visula files was also demonstrated. Unfortunately, circuit references were not included at that time and the manual addition of these was time consuming. An interface to the VAX network of some 60 terminals, used principally for software development, was offered but not demonstrated at this time.

One of the initial problems encountered in providing a justification for the system was that there were no figures for the cost of manually producing drawings, item lists and modification notes and their associated recording, filing, issuing, logging, maintenance and re-filing. A detailed analysis of DO, Technical Publications and print room activities on a time and motion basis was therefore carried out, and a reasonably reliable forecast of the minimum amount of time saved using DOMA, based on the existing workload, was produced (see Tables 10.1 and 10.2). This analysis also included basic drawing activities which brought to light that—although CAD might save 10–15 per cent on drawing time, the research time to produce a drawing was invariably at least four times the time taken to draw it, and so the total expected time saving would be unlikely to exceed 15 per cent of 25 per cent of the total time taken to produce a drawing, i.e. 3.75 per cent.

Engineering Data Management (EDM) would also make the data more readily accessible to assist in the production of drawings. Therefore, on the basis of drawing creation alone the EDM approach would produce significantly greater savings. However, there were other considerable advantages to introducing CAD, but they were dependent upon a good sophisticated EDM system already working.

The draughtsmen were initially unsympathetic to this approach as they felt that CAD was more directly related to the DO function. Drawing offices, however, have always been producers and distributors of production and purchasing information which embraces a wider function than only the production of drawings. Experience with EDM has changed the draughtsmen's view, and in spite of the frustrations of the inevitable teething problems they are now enthusiastic contributors to the development and thus the increasing success of EDM.

Subsequent visits from various interested companies endorsed RAVL's view that it is better to get the data control in place and working before attempting to increase what may already be a manually unmanageable data flow.

A proposal was produced showing the functionality benefits and the anticipated savings translated from time into cash. This gave the option to interpret the savings as the ability to take on more work for the same cost or maintain the existing workload at a lower cost. In fact it meant that the company would at last be able to cope with the existing workload and still have some capacity left either for extra work or accelerating existing work. In all of this it was necessary to produce a convincing case for management as an investment of around £350 000 was required. The original proposal envisaged a return on the capital within three years using DOMA.

The proposal included an ICL Ethernet network initially using 9 Sun 3/80 workstations, a Sun 3/480 file server, 4 Kokasai slave terminals, two nine-pin printers and two high-speed heavy duty printers. Three of the terminals with extended memories were sited in the laboratories for Visula schematic entry, one in the Production Engineering Department, and one in the Automatic Test Department where the network was also linked to a programmer and two MEMBRANE automatic PCB test machines. Bridges were also required to the VAX network and to the remote System 39 mainframe. The distribution of information was very much to the forefront of the proposal.

The system was overspecified for EDM only, but it was eventually expected to run CAD on the same hardware.

The data in OMAC was immaculate, calculated as being 97 per cent accurate, due to the database controller's previous intransigence with the OMAC entries. DOMA was therefore primarily loaded from OMAC in a few hours, with ICL assistance. However, several man-months of work were required to complete the item lists as OMAC did not contain specifications, circuit references, etc. This was undertaken by working back from the current production requirements and enabled RAVL to go profitably live in four months.

DOMA ran for about 12 months. Initially it ran well with two users loading the system to complete the data originally loaded from OMAC and entering the new item lists being produced by the draughtsmen. However, when the draughtsmen started to load the system themselves it slowed down, eventually to an unacceptable level, although even when running agonizingly slowly it was still profitable.

Unexpected problems included the complexity of data management, the addition of new users, the creation of new workpools and a developing instability as the system load increased. The CADCentre were engaged in rectifying this last point but it was felt that progress was too slow and that projected cost savings were being lost.

RAVL had, and continue to have, a very close liaison with ICL. ICL rapidly appreciated the problem and suggested a system based on Ingres which could replicate much of the DOMA functionality but would operate faster with simpler management procedures and a higher degree of reliability. RAVL requested a demonstration using a full database rather than a cut-down simplistic version and were impressed by the relatively intuitive screen presentations. ICL were prepared to offer the new system, called EMCS-X, on very reasonable terms as a replacement for DOMA. There were some initial problems about some of the longer reports but these were resolved before system delivery. EMCS-X also maintained the prime requirement that future developments could be undertaken in-house. The Engineering Change Request procedure was built in and there was a wider growth path for some of the developments that were planned (for example the print room distribution module, and the eventual absorption of in-house specifications).

The name 'EMA' (Engineering Management Avionics) was chosen to represent the RAVL version of EMCS-X, mainly due to the additional functionality which was subsequently added. It was about this time that ICL formed the Manufacturing Systems Portfolio Ltd (MSPL) division to manage, among other engineering software, EMCS-X.

With the introduction of EMA, the hardware was upgraded and a SPARC 2 machine was added as the EMA server, replacing the database controller's Sun 3/80. The 3/480 server is now being used for Visula files and archives. EMA currently occupies about 160 Mb and this is expected to double over the next three years as CAD and other documentation is taken on.

For EMA alone, the file server and database controller were felt to need the fastest hardware available for system development, special reports, and the monitoring and releasing of data to OMAC which could otherwise create a bottleneck. For creating item lists and the associated addition of data, probably relatively low level PCs could suffice. For accessing reports, 386 level PCs proved adequate, networked through PC-NFS or similar, and obviously have other uses than would dumb terminals. RAVL feel the choice of hardware may be more dependent on what other programs it is expected to run and therefore it is helpful if there is an overall IT strategy in place.

EMA had been running for about a year when another interface was required for FAMS, a new MSPL Factory Management System, to enable FAMS to take certain data such as pending changes from EMA. RAVL now have EMA talking to five separate nominally unrelated programs on six types of hardware, over three disparate local networks and two remote sites.

Many of the requirements which were unique to this implementation have been incorporated in later versions of EMCS-X. The database controller responds rapidly to requests and is creating new features at the rate of about two a week. He liaises with MSPL with a view to convincing them that certain of the new features should be incorporated into their standard EMCS-X system. The requirement to be able to adjust a system in-house to meet new requirements is considered paramount—RAVL consider it highly unlikely that a standard system will provide every requirement without some degree of change. The danger is that the new features built on may inhibit or affect existing features and may create problems and extra work with system upgrades. It is felt that the use of a standard RDBMS (Ingres) reduces this danger.

Unfortunately, the savings capable of estimation with the limited analytical resources available at the time were restricted mainly to the DO, print room and print library where the individual activities could be separated and timed with a reasonable degree of accuracy. This was not attempted for Racal Avionics as a whole (which would have been preferred), but a general appreciation indicated that the external savings overall were likely to exceed the DO and print room internal savings, assuming that the system was developed to its full potential *and fully used*. This last point was, and still is, important because the system may be unacceptable, particularly to shop floor users, for a variety of reasons. For example:

1   The system may not be seen to provide the information required.
2   The information required may be difficult to extract.
3   It may not be known that the required information is available.
4   The means of access may not have been explained (perhaps due to inadequate training).
5   There may not be enough terminals and printers or they may be wrongly situated.
6   A general fear of using computers at all may exist.
7   There may be a general unwillingness to make the system work.

RAVL feel that user acceptance is *essential*, and therefore the system *must* make jobs easier, and it must be seen to make jobs easier. The intricacies of the

EDM system sometimes manage to conceal this. Also, it must not be seen to *replace* their job.

Initially, draughtsmen complained that item list creation and modification involved too many screens, particularly when compared with DOMA. With more experience they accepted (with some reluctance) that the flexibility, control and reporting that this approach enabled made it worthwhile. However it is still believed that the on-screen DCF (Drawing Change Form) creation is overcomplex for casual users.

The benefits achieved

Savings

The savings shown below are principally the estimates made for the proposal. The automatic modification note was an initiative developed by the database controller, which initially was not believed could be achieved. Over three years the savings on that single function covered the total cost of the initial system.

During the manual loading of OMAC, because it was quicker to enter changes onto OMAC than change the paper item lists and RAVL were required to offer a 24 hour turn-round on bought-out parts, the paper item lists (which were the masters) were neglected and thus became out of step with OMAC. When DOMA was originally brought on-line there was an estimated two man-years of work outstanding. Transfer of the OMAC records onto DOMA left only a few man-months of work to complete the DOMA item lists, which proved to be a very worthwhile saving.

RAVL are currently operating with eight draughtsmen on mechanical design and two Visula draughtsmen, having lost six draughtsmen, section leaders and the chief draughtsman. An analysis of documents processed through the print room last year, compared with 1989, showed the throughput reduced only by 10–15 per cent with a 40 per cent reduction in draughtsmen and print room personnel. The backlog of over 2000 outstanding engineering changes has been reduced to around two hundred. The print room staff are enthusiastic users of the several control features and reports designed for or requested by them.

The authorization for issuing new and changed drawings is the 'batch file' containing all the items passed through by the database controller. On acceptance of this, one copy of each is printed off accompanied by its modification note defining and recording the last change, plus a list of all users with issue. The time saving on this is appreciable and nothing gets missed or issued without authorization.

It is interesting to note that when the original analysis was undertaken, as a percentage of time employed the potential savings were greater in the print room than in the DO. This is felt to be due to the print room spending more time in the distribution and control of information rather than in its creation.

Some of the savings made by EMA compared with the manual system
Total savings have not been made from a few large unit savings, but from a multiplicity of small savings on frequently repeated tasks. These small tasks will come up in an analysis of how a company functions, and aggregated costs are the result of how many times the tasks are repeated. As many of these small tasks as possible should be programmed into the system. Savings come from the number of manual tasks which the system either expedites or removes. Many of the reports now available in EMA would not have been considered as being available from a manual system because of the time involved, or would not have been possible because supporting data was lacking, particularly where that data is itself a function of other data associated by EMA, for example DCFs (drawing changes) outstanding on a project for a specified period.

Some of the savings shown in Tables 10.1 and 10.2 have not been directly translated into cash as it has not always been possible to estimate the total volume of demand with sufficient accuracy and would vary from site to site or company to company. Although the estimates were made for DOMA, the slowness of the system did not enable all of them to be realized. However EMA has achieved these savings and in many cases exceeded them. It should be noted that many users have considered the manual estimates optimistic. In general, however, the savings are regarded as the minimum expected (i.e. not 'hyped-up').

The savings are a function of the throughput of the DO and would be greater with a larger proportion of new work rather than variants, since the generation of completely new item lists would remove the problems occasioned by the necessity of correcting old paper item lists which frequently either omit data or contain inadequate or obsolete data which EMA will not accept.

Shop floor saving example   Going to the print library from the shop floor, requesting and signing for a print, waiting for service and returning to the shop floor, takes about 15 minutes or longer. At the last analysis in March 1993 there were approximately 200 requests per week. If the required item lists and drawings were available from a terminal and printer/plotter at the point of use, the savings would be about £60 000 per annum, or a shortened production time about equivalent to employing two extra personnel.

Other current benefits of EMA
1 Instant availability of documentation 'state' to shop floor. This previously required a visit to the print room or a phone call.

Table 10.1 Savings made by EMA

		EMA	Manual	Saving
1	Typical item list creation			20%
	Typical item list mod			50%
2	Drawing list	On demand	2 days to 3 weeks	2 days to 3 weeks
3	Master record index (this will be on demand when the system is complete)	2 hours	4 days to 3 months	4 days to 3 months
4	Modification note	On demand	30 mins	30 mins
5	Specific part type usage	On demand	1 hour	1 hour
6	Where used	On demand	Not available	(Available on OMAC)
7	Family tree	On demand	1 day	1 day
8	Planning tree (creation)	3 hours	3 hours	None
9	DCFs outstanding on drawing	On demand (accurate)	10 mins (based on print room interpretation)	
10	DCFs outstanding on section	On demand (accurate)	30 mins (based on print room interpretation)	
11	DCFs outstanding on DO	On demand	At least 7 days manually, or 3 days using Delta data	(Note: Delta is a PC-based database which preceded DOMA)
12	DCFs Outstanding on project	On demand	20 mins if available in section leader's file	

2 Automatic re-formatting of item lists for inclusion in handbooks and manuals. This was previously subcontracted at extra cost and delays.

3 Outstanding changes to item lists and drawings are listed at the head of each item list providing a quick check that the changes have or have not been incorporated, and that they have received all the relevant DCFs for changes not incorporated.

4 Prints held by users are listed on EMA as part of the print distribution module and may be accessed by users with permission. They are sent a paper copy once a week to remind them of what they hold and to return any document no longer required.

5 Remote sites no longer require prints or microfilms of item lists, other than for servicing purposes for old equipment (RAVL still service equipment 30 years old). On DOMA RAVL were able to print off any issue of a single-level item list, but this feature has only recently appeared on EMA.

Table 10.2 Specific examples of savings

		EMA	*Manual*	*Saving*
1	Multilevel item list for DVOR (E524 items with 588 assemblies)	3 mins (for 5700 items +30 min printing time	Not available	
2	Family tree for DVOR	3 mins + printing	2 days	2 days
3	DO DCF analysis	10 secs	1 week manual, or 3 days using Delta data	
4	Item list mod notes (creation)—creation automatic and accurate on DOMA	25 mins (including checking time)		Approximate savings around £160 000 per annum. (Based on 6 I/L mod notes per day per draughtsman for 9 draughtsmen, including time required for checking on manual system)
5	Mod note retrieval time	On demand		1 min in print room. (Estimation of 40 000+ retrievals per year pre-DOMA saves about £10 000 per annum. But could be 3/4 times this pre-DOMA as accurate figures are not known)
6	Mod note print time (current mod note automatically appended to foot of I/L)	1 off per I/L required. 5 secs		1 off per copy of each sheet pre-DOMA. Estimated 10 000 sheets per month (10 secs per sheet) saves around £10 500 per annum
7	Item list retrieval time (based on 10 000 sheets per month = ave of 5000 item lists)	On demand	3 mins per I/L	£40 000 per annum.
8	Notifying the print room of I/Ls and drawings to be issued	Automatic via batch file	5 mins per bundle in print room	£330 per annum
			5 mins per bundle in DO (Approx 270 bundles annually)	£660 per annum
9	Reduction in microfilm requirement.			£12 400 per annum
	Initial OMAC database adoption			£75 000

6   The print issue record held on 'white cards' showed an unbelievable 14 000 prints issued for a company about 500 strong. The records now checked and held on EMA show about 1500 on the shop floor at any one time. This has relieved the print maintenance problem and it is proposed to make all data accessible at terminals and printers on the network thus placing the responsibility for ensuring that the latest issues are being used onto the actual user or the section supervisor (since the issue can be checked from EMA in seconds).

7   EMA has many of the DO procedural requirements built in. Checking requirements are therefore reduced, and, when the total system embraces drawings with a similar level of control, there will be less need for draughtsmen to be concentrated in an office. They could be moved to any point on the network convenient to the project development group, which could be a remote site or even at home via dedicated lines or modems.

System details and points to consider

The following points are suggested by Racal Avionics Limited as those which should be considered when selecting, implementing and using an EDM system:

1   Before considering a system, analyse how you actually work, determine how you would prefer to work and produce your own *Functional Requirement Specification* (FRS). This is likely to be appreciably longer than you originally thought. Break each requirement into its component pieces as each piece needs to be programmed and if you leave the analysis to a programmer he or she may miss some of the subtleties and it may be difficult for him or her to appreciate the value of the function. Consider how you may wish to expand in the future, and the requirement for external links to other packages for data exchange. RAVL felt they were very fortunate in that the ICL programmer quickly appreciated both the utility and the value of the functions that were requested and in many cases improved on them. Where company personnel have inadequate time or expertise to conduct the necessary analysis, it would be cost effective to employ an outside consultancy to ensure, as far as possible, that your system foundation (i.e. the FRS) is broad-based, far seeing, and not a hastily 'cobbled together' wish list with gaping organizational holes which may preclude its economic development.

2   Select a system which meets most of your criteria. If you find one which meets all of your requirements, you probably have omitted something from your FRS.

3   Keep your working committee as small as possible but advise other areas of what you are doing to avoid conflict with whatever they may have in mind. The working committee set up by RAVL consisted of three people.

Their tasks included: to produce the FRS, to find and evaluate the systems, to select one and propose it, and eventually to load the system, write the operators' manuals, train and advise and then to run and develop the system in line with internal requirements and user requests.

Large, fully representative committees get bogged down in endless meetings and dilution of essential functions. Such committees are frequently overloaded with managerial personnel who will not be using the system 'hands on' and have traditionally relied on other people to produce the data they require. If anyone else is required on the committee, choose people who actually create data, seek it, and use it. However it helps if you can establish a sympathetic relationship with someone with financial clout.

RAVL's own committee jointly had appreciable experience of the DO problems, experience of the shop floor, buying and stores procedures and appreciated many of the problems which were manually insolvable with limited human resources. As each requirement was identified, a system solution was noted, even if it lacked elegance and subsequently was revised in consultation with the programmer. Investigating systems enabled them to clarify their view of what was achievable.

4   The selection committee *must* be enthusiastic and determined to achieve results. Implementation is not a 9–5 job which can be switched off in an instant—it is more a way of life. As with any design function, ideas come at strange times.

5   Do not accept a lower priced system which will not enable the major savings or future developments to be achieved. If it does not meet expectations credibility will be destroyed, there will be user frustration and there will be even more difficulty in funding future developments.

6   Although it may be cost effective to put the whole planned network in at an early stage, do not be tempted to go live throughout the company simultaneously other than for 'read only' reports. Get your database working and loaded to a useful size; try to avoid half the data on paper and the rest on the computer. There will inevitably be problems which may be difficult to isolate and will destroy user confidence if not rapidly corrected. Get a local pilot scheme working properly so that you can at least demonstrate what it was supposed to work like before extending into possibly unfriendly territory.

7   At the initial phase ensure that you have recruited a strong minded database controller who will manage and administer the system while resisting suggestions to corrupt or change it. If the system will not initially accommodate a necessary change or modification, a program should be devised which will achieve the desired result, but probably in an unfamiliar way. The database controller should agree the new system and either personally devise the programs or arrange with the supplier for the required

modification. The database controller is the person most likely to appreciate the knock-on effects of readjusting the system.

8 Remember that the final acceptance of a system depends on user satisfaction. To achieve this, the system operation must be as simple as possible and it must be apparent to the user that it has solved a problem. This simplicity may require that the programs are complex, but do not accept a programmer's view that it would be preferable for the user to learn more about the system. Remember the programmer writes the program only once (with any luck), whereas the user may be using it many times an hour. Listening to the user and embodying his suggestions where possible helps the user to appreciate that it is 'our' system rather than 'your' system. The user does not appreciate the elegance of the programming, only the ease of use.

9 There is a strong probability that the system proposals will be received with restrained enthusiasm by the management, bearing in mind the appreciable investment required. The return on capital is likely to be higher than almost any alternative investment. It will in many cases, assuming an adequate workload, repay the investment in two years or less, and continue to repay in ensuing years as its uses become more fully exploited and its functionality is developed. In depressed climates, however, cash may be needed for more immediate debt repayment etc. Unfortunately, if the system is delayed until the workload is assured beyond doubt, greater disruption will occur—possibly at a time when the company can least tolerate it, but at a time when the savings would have enabled a stronger presence in a competitive market. Consider scaling down to a properly representative pilot scheme which, when working, could be expanded relatively painlessly. Do not settle for no scheme at all—you will be losing money.

10 It is important that every effort is made to ensure that the data loaded is correct. Circulating errors more efficiently also increases the cost of rectifying them.

11 From the outset RAVL's systems were never considered isolated operations within the DO. The interfacing with outside systems and the distribution of data around the company was always a guiding consideration.

It was not made easier by the lack of a documented company IT strategy. An analysis was undertaken to show what data was required from which sources and who it was felt should use the data produced. Even this limited approach showed the value of easy data circulation and access. The lack of a cohesive IT strategy remains an inhibition in that RAVL can never be sure that what they do will provide data in the form that other systems will eventually require or where, in the long run, it is best to source the data. It is preferable that areas generating basic information are responsible for the loading and maintenance of that information, and it is even better if that area also suffers the consequences if the data is erroneous or corrupted as it provides a spur to get data right first time. To enable this, the system must control access for data entry and correction to those specific areas of responsibility in order to avoid accidental corruption of other people's data.

12 Taking control of external documentation such as test and performance specifications (which are currently held in three word processors on three types of hardware) and drawing office procedures and specifications (currently held largely on paper) have problems depending on the degree of control expected. For total control an appreciable amount of formatting effort is required and funding for this area is planned for the future.

However, when all production data is accessed through EMA it will give an opportunity to re-evaluate the way the company actually functions. So much of a company organization is structured around the operational control by paper with its attendant signatures, circulation (probably several times through several levels), postal records, departmental incoming document recording, copying, receipt signing and filing of the same document in many locations. This generally results in loss of documents, the forgetting to action them, etc. Most of this can be handled with automatic recording and transmission by an electronic data handling system, with no document loss, alerting recipients that the document has not received attention and with the whole distribution taking only seconds. With EMA, pending changes are automatically directed to the responsible section leader's 'workpool' which is a safe area to which only he or she has 'write' access, but can be viewed by anyone with suitable permissions.

13 The prime contact a user has with the properties of a system are through 'help' screens. Initially RAVL suffered greatly by them not being available. Help screens as supplied by vendors have a tendency to contain messages such as 'error 9534/axy' which, as there is usually no way of finding the hidden meaning which was probably only intended to help the programmer and not the user, succeeds in being useless. Help screens *must* be in-house configurable. This enables the database controller or system developer to respond to users' actual problems as they occur. Users' problems are seldom unique and an on-screen solution would be helpful to all. System designers are not the best people to make monolithic decisions about users' problems.

Items in a menu frequently cannot adequately describe the related content detail. A Help screen attached to each menu item with a full explanation of the information contained in that item and relevant cross-references would lead to a greater understanding of the system's functionality around a wider audience than a paper manual is likely to achieve, particularly among casual users who have little opportunity to gain experience with the system.

A further 'User Guide' on screen, based on a question and answer approach could steer users towards the programs they need as opposed to working through the 80+ menu options to find what they want.

RAVL feel that it is worth bearing in mind that, for example, whereas draughtsmen are relatively intense EMA users, using about 30 per cent of the functionality about 40 per cent of the working day, most external users will use only about 5 per cent of the total functionality for about 2 per cent of their working day. These users will thus have little opportunity to explore beyond their immediate, familiar requirements and may still be accessing other data the hard manual way, not realizing that EMA has the data they require instantly accessible.

14 There is difficulty in keeping people informed as to what is being done and what the use of it is. Newsletters have been issued whenever it was felt there was a significant advance—these also included future plans. Sixty copies were issued with instructions to circulate them. Unfortunately, evidence suggests that if they are read their importance is not appreciated. Casual conversation can produce responses such as, 'I didn't know that was available—it would have saved me hours of work.' The information was contained in the newsletter that had been received, but had not been read or assimilated. A copy of the latest menu is always included and an exhortation not to undertake any manual listing without first consulting IT. The possibility of explaining one menu function per day on the entry screen is being considered. Maximum system advantage will not be gained until each user questions why they have picked up a pencil and are looking for a piece of paper.

15 Budget on the basis of a company-based system. In RAVL's case, potential major savings have not been achieved in other areas due to the lack of the necessary hardware. With a central budget the facilities can be provided company-wide and, once they are available, people can be persuaded to use them, experiment with them, and develop an appreciation of the benefits. In this case, it was the introduction of the Factory Management System and its union with the existing network which made EMA more widely available (although not on an organized 'need to know' basis), and has helped in an appreciation of what can be achieved for the individual and thus the company by making more use of EMA data.

16 Do not underestimate the capabilities of staff to operate the system. RAVL initially worried that the staff in the print room, with no keyboard or computer experience, would experience difficulty in adjusting. Fortunately the reverse was the case. There was about a three-day learning

curve before they started requesting additional facilities, and have continued to be among the most enthusiastic users.

17 When selecting personnel either to administer, operate or document the system, spread the 'net' widely. More computer literate people may exist than are realized and more people comfortable with the psychology of computers who can rapidly assimilate the requirements than is obvious. Some older managers are not comfortable with computers, may think that their subordinates are similarly afflicted, and thus be reluctant to offer their services. It is frequently difficult for an employee doing a good job to persuade his boss to lose him to a better or more interesting job as he would be difficult to replace.

18 User manuals describing each function on a keystroke by keystroke and resultant screen basis had to be written. These have been maintained, adding user instructions for each new development. With the advent of BS5750 creative users are reluctant to use a new procedure until it is documented. Producing and maintaining these manuals is initially a full-time job and if the developing procedures are to be produced within an hour or two of the new procedure being required (which is a frequent requirement as the new procedure is often the result of an emergency), someone must be available who can drop everything and concentrate on the new requirement.

19 RAVL managed to load the system, write and maintain the manuals and procedures, test the system, develop new programs, and generally manage and administer EMA with two people. This was only possible because the database controller is a workaholic who starts at 06.00 hours when nobody else is on the system. There is no way of inhibiting users and certain maintenance tasks can only be undertaken when the system is not in use. Similarly it is better to try out new programs when the system is not being used. Allowance must be made for this when setting up the parameters for a system management team. The testing of the system and the writing of the manuals was done simultaneously by the same person. When system errors were found that portion of the manual had to be temporarily abandoned until a solution was found. Meanwhile work continued on another section of the manual.

User comments

RAVL feel their users tend to be either outspoken or sit in a corner muttering, a practice which is discouraged. They are encouraged to note their comments on an 'EMA Development Form' if outside to the DO, or entered into a log if within the DO. Comments vary from, 'The system is a pain but I wouldn't be without it' (draughtsmen) to 'It's made my job much easier and more practical. The data is so easy and quick that better situation evaluation can be made.' (report user).

There is a general impression that as the system develops it is appearing to become more user-friendly, but facilities such as the ability for a user to configure his own menus would be appreciated.

Conclusions and future aspects

RAVL consider they still have a long way to go to control all associated engineering documentation, i.e. drawing office specifications, company procedures, and drawings. The planning function in the drawing office is particularly difficult for their small batch, diverse product, fluid situation without spending a disproportionate amount of time on the actual planning using very limited resources. Although active draughting effort has reduced by about a third, an analysis undertaken in March 1993 showed that the prints issued reduced by only 15 per cent as far as could be checked manually. This could be improved by the use of CAD. Sometimes the drawing office is bypassed in the mistaken belief that drawings could catch up with sketchy inaccurate production information at a later date. Although the belief that the drawing office could not meet the undefined timescales was erroneous, there is nonetheless a higher level of confidence in the ability of CAD to deliver on time.

RAVL now have a program to check on the quantity of new and modified drawings and item lists produced by the drawing office and also the quantity of prints issued from the print room for any stated period and are therefore in a stronger position to evaluate the effect of any future developments on DO output.

Short Brothers plc

Introduction

Short Brothers like to play a leading role in aviation technology. From making six 'Flyers' for the Wright Brothers in 1908, to the design and manufacture of the world's first vertical take-off and landing aircraft in the 1950s, the company has a tradition of innovation. Shorts are now part of Bombardier, a Canadian Corporation engaged in design, development, manufacturing and marketing activities in the fields of transportation equipment, motorized consumer products, aerospace and defence.

The company, which employs over 10 000 people worldwide, has its headquarters in Belfast, Northern Ireland, with subsidiary offices in London, Washington, Kuala Lumpur and Bahrain. Since privatization in 1989, Shorts have been transformed through a £200m investment programme in plant, machinery and facilities. Shorts are now a centre of excellence for design and manufacture of Nacelle Systems, fuselages, flight controls, close air defence systems and for processes such as composites, metal bonding and computer-aided design and manufacturing.

The company has a clear strategy of growth in both aerospace and defence. This includes maximizing its European presence through joint ventures and risk-sharing partnerships. These include Hurel-Dubois in Nacelle Systems, Thomson CSF in close air defence systems, and Fokker/DASA in the Fokker 100/70 programmes, and developing relationships within Bombardier with Canadair, Learjet and de Havilland, thereby increasing capabilities. Joint projects are undertaken requiring the sharing of engineering information at all stages of design and manufacture. Control of engineering design data and the control of the design process is therefore vital. Shorts have recognized the benefits that EDM can provide in this environment and this case study describes some of the issues and the approach taken in the implementation of Computervision's EDM system.

Background information—the need for EDM

Short Brothers' first investment in CAD was the installation of seven Computervision CADDS3 workstation terminals in 1981. Over the intervening years the number of workstations has grown to a current total of around 250 (230 CADDS and 20 CATIA), located at the company's various sites. An additional 27 workstations running Formtek software have been installed for the distribution of drawing data in raster format. The Formtek files, which are generated by scanning microfilm copies of existing manual drawings or by direct conversion of Vector CAD files, are also used for illustrated process planning. For the first eight years of operation, the CAD system was only used as an electronic drawing board, but in 1989 a team was formed to spearhead the integration of design to manufacture. This has resulted in the current comprehensive Computervision (CV) installation. However, until comparatively recently, the management of CAD-generated data relied upon a number of small independent servers, with individuals filing over the network. Security backups were performed at each individual server, with archive data being filed to tape.

As the size of the network grew a number of problems with this approach to data management became apparent. For example:

- A single failure at a server level would disable a number of CAD workstations.
- As the physical size and complexity of parts increased, filing across the network became a serious bottleneck.
- There was no continuity checking between the servers with the consequent possibility of the same part existing at different revisions in different locations.

It was envisaged that these problems would be further compounded by the following design trends as identified by Shorts:

- Further increases in part complexity
- Electronic mock-up
- Reduced development times
- The manufacturing requirement for CAD data

It was therefore felt necessary to introduce some form of data management system to ensure that Shorts could move forward and accommodate these changes in the engineering process, unhindered by data management concerns. The engineering department therefore took control of this need and has driven the solution to its current form.

Short Brothers identified a number of areas as being the major driving forces behind their need for EDM:

- *Quantity of data* The quantity of design information at Shorts is vast. A typical aircraft contains around 100000 to 150000 distinct parts. Although some of these will be standard parts and thus not require design models or drawings, the majority of parts require design drawings, assembly drawings and data, stress calculations, test data, manufacturing data, and so on. Visibility, accessibility and control of this data, and all their associated changes, is essential.

- *Data availability* Before EDM was implemented, a 'Design Index' system was available to report what design files existed and their associated revision. However, this was a reporting only system and provided no control over the access to and distribution of design files. The data in the system was entered manually and therefore lagged behind the real status of drawings. The process of obtaining access to released design data files for viewing or for modification could take from hours to days. Immediate, controlled access to all design data for various contracts and from the defined standards was required.

- *Remote sites and subcontractors.* When designs were subcontracted to other sites (whether internal Bombardier or external to the group), access to existing or related design files was a greater problem still. In addition, visibility of the progress being made on the work at these sites was difficult to obtain.

- *Improved design workflow* The design and manufacturing process at Shorts used to be controlled by a series of paper-based forms. Changes were initiated and authorized by completing and distributing paper forms, and design drawings could not pass from one stage to the next without circulating several pieces of paper. As might be expected, forms were incorrectly completed, got lost, left in someone's in-tray, and so on—all resulting in delays to the final product. These processes needed to be automated, progress made visible, any bottlenecks or delays immediately highlighted and errors minimized. An Engineering Process Innovation Centre (EPIC) was therefore set up to re-define the overall process and to implement the necessary step change. As a result of this, Design Build Teams (DBTs) were introduced and the concept of EDM was initiated.

- *Support of Concurrent Assembly Mock-Up* Shorts have been awarded the contract to design and build the new Learjet 45 executive jet and are responsible for the design and manufacture of the complete fuselage and empennage. The time available for the design and manufacture of this jet is a small fraction of that previously required for similar projects and, as a consequence, the latest computer aids are being used to address these issues. The main technology being used in this area is 'Concurrent Assembly Mock-Up' (CAMU). This allows entire assemblies to be generated on computer and checks such as clash detection performed. This environment requires EDM to manage the volume of design, assembly and configuration data which needs to be accessed simultaneously by many designers.

The key aims of the EDM solution (which was implemented at Shorts in several phases) were identified as follows:

- To provide fast, accurate, controlled access to design data
- To interface with the existing Design Index and Process Planning systems
- To improve the design process by speeding up the transfer of information, thus increasing the visibility of design progress
- To provide an environment to support CAMU
- To implement configuration control through EDM
- To provide data access to remote Shorts sites and to subcontractors

This functionality needed to be implemented in such a way as to cause minimum disruption to the designers, particularly since the Learjet 45 contract was already starting at this time. This meant it needed to use existing Shorts terminology, run on existing hardware and require minimal training. Clearly, the designers required a system with an easy-to-use GUI tailored to their specific needs. Some further customization was also required to interface with the existing Design Index and CAPP systems.

System selection and implementation

For some years now, Shorts have been investigating the various data management systems available with a view to implementing a corporate data management system. The IT department, which conducted this review, concluded that there was, as yet, no system which fulfilled the corporate requirements and that the investigations should continue. Simultaneously, development work within Shorts had resulted in the 'Design Index'. The Design Index was a database system which maintained information on all drawings, drawing sheets, their current status and revisions, etc. An 'optical juke box' was installed to hold the archived CAD and raster data, indexed via the Design Index.

A joint decision was taken by the IT and engineering departments to pursue a CAD-centred data management system. A review of the available options highlighted Computervision's EDM as being the most promising, for the following reasons:

- Already proven functionality in the required applications
- Good working relationship with the vendor
- Close integration with the CAD system was likely to be maintained through future developments (e.g. Mock-up, Parametrics)

A decision was taken to obtain a demonstration copy of the EDM software and a six month pilot project, involving approximately 12 different users, was undertaken. This pilot confirmed the suitability of EDM for Shorts requirements and a full-scale implementation was planned.

With respect to hardware in particular, a DEC VAX mini was provided, in line with the then current company IT strategy, for the EDM implementation. During the early implementation stages it became clear that the available machine would require considerable upgrading to achieve the required performance level and the opportunity was taken at this time to move to a UNIX environment with the substitution of a Sun MP690 Model 140. The Sun server, originally configured with 23 Gb of disk and 128 Mb of RAM has since been upgraded with an additional 20 Gb of disk. Additionally a 32 Gb optical juke box introduced into the VMS environment for on-line archiving (replacing tape archiving) has been transferred to the Sun server.

To date, the timescale of EDM implementation has been:

August 1992	Initiate EDM system specification.
November 1992	Install first phase of EDM system running on Sun workstations. Approximately 100 users on a single site controlling 10 Gb of CADDS and raster design data. Interfaces with Design Index and CAPP. Approximately 2 hours training per designer.
May 1993	By this time approximately 200 users controlling in excess of 50 Gb of data. Remote sites accessing data over ISDN links. Support of CAMU environment. Data fed into project management system (Artemis) running on PCs in addition to Sun workstations.
June 1993	Subcontractors have direct, controlled access to data. Product structure held in EDM, supplying reports on assemblies. Configuration control implemented.
Quarter 3 1993	Compound documents. Enhanced change control.

The volume of data being stored in EDM is still increasing rapidly as more sites are linked into the system (Learjet themselves having a direct link now) and more types of documents are included. EDM is now the corporate data store for most engineering (and some non-engineering) data.

The benefits achieved

Ease of use
One major aim of the first phase was to provide fast and easy access to 10 Gb of CAD drawing and model files. The designers would then be able to access these model files for 'view only' or for 'modification' in a matter of seconds instead of the hours or days that it took before EDM.

Design engineers were used to computer systems providing a graphics-based user interface, with mouse-driven windows, icons, scroll bars, and so on. This technology was also required from the EDM implementation. A file access user interface was therefore developed in conjunction with the users, and has resulted in a system that has been universally accepted from the very first pilot system. The interface is very intuitive to use and system upgrades do not require additional training effort.

Control and availability of design data
The EDM system not only provides fast, controlled access to information related to both projects and company standards, but it also controls the status of drawings/models under its control. It therefore knows when drawings are created, when they are raised in revision, and when they are released. This type of data, and the associated reports, can be compared against the outstanding engineering change requests to provide reports such as how many changes are outstanding, what changes are taking longer than a specified number of days to complete, etc. Prior to EDM, obtaining this sort of information was a purely manual process.

Shorts feel it is almost impossible to quantify the savings in time and effort brought about by having information and drawings immediately available and knowing that the information is correct. The real measure of improvement will only be seen when full production begins—only at that stage will it be known if extra access has actually improved the design and resulted in fewer production modifications. Certainly the CAD department could not have coped with the increasing volumes of data as a result of the Learjet 45 contract without the use of EDM. Within about six months the volume of data held in EDM has increased from 10 Gb to over 50 Gb and no increase in the CAD support team has been necessary from the initial implementation.

Attributes associated with CADDS parts
Each CV CADDS drawing or model has many attributes associated with it. Some of these attributes are automatically controlled by EDM, such as re-

vision, date last accessed, name of user; others (such as effectivity, weight, dash number) are user controlled. Traditionally, these attributes were held on the drawing itself, so anyone wishing to access them needed to access the drawing—a time-consuming activity. With EDM they are now held as metadata, outside the CAD file, and under the EDM system's Oracle database. This provides a greater degree of controlled access to such information, which Shorts believe is of great benefit.

Product structure/configuration control
In any product the BOM is one of the most important pieces of design information. This assembly information is created using Computervision's Assembly Design software package. The assembly information is stored and updated into the EDM system. Each node on the assembly may have a specific CAD file associated with it. This CAD file will exist in EDM and have the attributes as described above associated with it. Each level of assembly and sub-assembly can also have attributes associated with it, such as revision and effectivity. By associating effectivity with each CAD part and with each sub-assembly, Shorts have implemented their own configuration control system. Thus the design data for any aircraft can be regenerated as required.

Specific mock-up
By using the configuration control aspects mentioned above, the EDM system will either list or copy to local disk all the correct revisions of CAD data for parts that make up a specific assembly. Therefore, if a user wants to re-create a specific build of any assembly, they simply identify the assembly name and effectivity required. The appropriate CAD files are then copied to disk (assuming enough space exists) and the Assembly Design software can be used to graphically re-create the assembly.

Concurrent Assembly Mock-Up
Designing an entire aircraft (even without the wings as for the Learjet 45) is a substantial project. The CAD software used at Shorts allows their designers to ensure that their parts fit perfectly alongside the parts of others, thus promoting the use of concurrent engineering techniques. Each designer has their own personal area of disk space that they use to modify parts. When they complete a section of their modification, or after half a day's work, they update the changes back into EDM. This operation performs two main purposes: it saves their changes onto another area of disk which is backed up by the system administrator, and it also makes those changes visible to other designers. The update operation also takes a copy of the changed part and places it in a

central 'read only' area. The parts in this area are used to generate the 'mock-up'. A vital part of this system configuration is that EDM ensures that the read only area always contains the latest version of all parts. Several 'read only' areas exist for the Learjet contract, and several areas also exist for each zone of the aircraft, since the majority of designers only work in one specific zone. However, certain designers or managers have access to as many of these areas as required, and can therefore generate a 3D model of as large a section of the aircraft as required.

With the use of leading-edge CAD software such as EDM and Assembly Design, Shorts feel that electronic mock-ups are finding and eliminating design faults which would only have been detected at much later stages such as physical mock-up and, more likely, in production itself.

System details and points to consider

The EDM system controls the following categories of WIP, released CAD, and raster data:

CADDS files Geometry files generated using Computervision CADDS software.
Raster files Geometry files generated by scanning of manual drawings or direct conversion of CADDS files.
Release and change notes Raster files generated by scanning the release and change notes produced for each modification.
Illustrated process plans Data for process planning based upon the modification of a geometry raster file.
Drawing attribute data Data such as latest issue, status of a drawing, etc. Currently held in the Design Index.

User Types Analysis of the system users accessing data highlighted four main departments. The data accessed by each of these departments is summarized as:

Production	Raster files
	Release and change notes
	Drawing attribute data
Process planning	Raster files
	Illustrated process plans
	Drawing attribute data
Design	CADDS files
	Drawing attribute data
Reprographics	Raster files
	Release and change notes
	Drawing attribute data

Drawing file types   Drawings produced at Shorts can be categorized into ten different types, with each file type having an associated life-cycle. They are:

STD	Standard parts
LYT	Loft layouts
DRG	Drawings for issue
TDR	Drawings for issue (tool design)
SCH	Schemes
DEV	Miscellaneous data required to be archived
TEM	Temporary, user specific data, not to be archived
ADT	Assembly files from assembly design
PPS	Process plan sketches
IMG	Rasterized drawings, change and release notes

The design process   During the initial stages of EDM implementation it was not intended that the system would provide a high degree of integration and/or management with the evolving engineering processes. However, it was felt that some additional file control and management could be provided within the current working practices of drawing review and approval.

Analysis of the design process within Shorts revealed variations across different contracts and divisions. There were, however, many common stages in the process. These were taken and used as the basis for the definition of the overall design and detailing process.

Projects   Projects within EDM provide a grouping mechanism to control user access to files and to define a specific lifecycle to a group of files. Separate projects are used for each of the defined file types since each has a different associated review cycle. Projects are also dependent on contracts or departments in order to control access levels to data.

Thus each EDM project within Shorts consists of a single contract or department and a single file type.

Revision sequences   The revision sequences in use at Shorts are (apart from their own) generally those imposed by their customers. There are currently four revision sequences in use on different contracts.

Command lists   There are eight types of user identified on the EDM system used at Shorts: read only users, reviewers, administrators, and so on. Users can have a different command list per project.

Users   In general, each EDM user has their own user ID and password. This is a three-digit code currently in use throughout the design departments.

Naming conventions   A rigid naming convention is applied to all parts stored within the Shorts EDM system. The structure of the convention varies between the different file types and across different contracts, but the first three letters always denote the file type, e.g. DRG or SCH. The rest of the name denotes the contract name and the individual part identifier. Additional fields can be incorporated to identify assembly zones, sheet type and sheet number. Within EDM the part revision is held as a separate attribute and not included in the part name.

Integration aspects

Two major considerations when developing the user interface were that it be graphical in nature and that it be as standard as possible over the various departments and applications. Computervision's Accessware software was chosen to develop the graphical user interface because it had already been used for some EDM integration and it ran on all the necessary platforms.

Keeping the look and feel of the GUI the same across all application areas was felt to be very important. At the EDM sign-on level this resulted in the same logon interface for all users. The various application-dependent interfaces are outlined as follows:

- *CADDS integration* Having signed on to EDM using the standard menu, CADDS users (primarily designers) are then able to perform a wide range of file access and administration tasks associated with the design and review processes.
- *Assembly design integration* A default level of integration is supplied which allows the required CADDS parts to be accessed via EDM. The user specifies the AD tree required and EDM displays all the metadata associated with the assembly. The assembly can then be booked out/in etc.
- *Raster file integration* The raster interface is subdivided into the two functions of Production (view only) and Process Planning:
  - Production use of EDM is primarily aimed at obtaining fast local copies of drawings. Sign-on to EDM is once again via the standard interface with the user then entering details of the drawing required. Where a complete specification of the required data is not given the user will be presented with a list for final selection. The selected drawing can then be viewed and/or printed locally.
  - In addition to the production functionality Process Planning users can create sketches from existing raster drawings and store them in EDM.
- *Reprographics interface* When a completed drawing has been checked and booked into the Design Index, EDM is automatically updated and the CAD geometry file moved to 'released' status.

  The Reprographics department also has responsibility for the booking into EDM of scanned manual drawings, release notes and change notes.
- *Design Index integration* Communication between the Design Index and EDM is made via the transfer of data files in a batch process run every hour.

The Design Index informs EDM of all drawings successfully logged by sending drawing number, sheet number, sheet type and revision. This triggers the archival of the drawing concerned and the generation of a raster file.

When a raster drawing has been generated (by scanning or conversion) and stored in EDM, metadata relating to the file is transferred to the Design Index. Similarly metadata concerning any change notes or release notes stored in EDM is transferred to the Design Index.

- *CAPP integration*   Integration with the CAPP system is currently limited to the use of a data file which is already created by CAPP. When a process plan is moved between accounts a data file is produced with details of the move. EDM will read these files and when a file is moved into account 'EPR', any associated process plan sketches in EDM are raised to 'released' status.

- *Data loading*   At the start of the EDM project there was approximately 7 Gb of data on the optical juke box (under the VMS environment). In addition to the data itself, a significant amount of metadata was also stored in the Design Index.

   During the first phase, all of this data (and the metadata), was transferred from the OJB and VAX to the Sun environment and stored on magnetic disk. The transfer was effected using a series of scripts running over the DECNet Link to automate the task.

   There was sufficient disk space available on the Sun to allow the archive data to be held on-line during the second phase of development, at the end of which it was moved to the optical juke box. This procedure allowed a time window for connecting the optical juke box to the Sun and testing its integration with EDM before putting it into service.

EDM administration

Initially standard EDM backup utilities are being used to back up each storage pool onto a 2.5 Gb Exabyte tape. Eventually it is envisaged that backups will be automated using 10 slot Exabyte 5 Gb stacker units. Two of these units will be used in parallel to provide dual backup copies. One copy will be stored locally in a fireproof safe, the other being removed to an off-site storage facility.

Training

All training was carried out at Shorts own training facilities via a series of customized courses:

- *EDM admin/programming*   A 5 day course covering all aspects of the

EDM implementation such as backups, archives, installation and programming.

- *EDM administration*   Phase 1, two days and Phase 2, one day. This covered basic administration functions associated with the day-to-day running of the EDM system.
- *User courses*   Training of approximately half a day was given to each end user. Each course was customized to suit the needs of the individual types of end user, such as designers, planners, approvers, reprographics.

Conclusions and future aspects

Several areas for further development have been identified by Shorts. These include the automatic generation of an 'as designed' bill of materials and closer integration with the project planning and MRP systems. Developments in these areas will need to be kept in line with the requirements expected of a corporate data management system.

There can be no doubt that the EDM implementation at Shorts has been a resounding success. The system has met with universal acceptance from its users, and with those not included in the first phase lobbying for early inclusion in the second phase. It has provided the design supervisors and project managers with a much greater insight into the progress of work within their departments via the automatic reporting of completed drawings.

Significant benefits have also been realized on the systems management front. The reduction in workload has facilitated a 50 per cent increase in the number of installed workstations without any increase in operational staff. This increase in workstation numbers is primarily due to the increase in the complexity of the data relationships associated with the continually improving engineering processes employed at Short Brothers.

However, the greatest benefits are just starting to be realized via the widespread adoption and support of Concurrent Assembly Mock-Up in the engineering design environment. Shorts feel that the management of the vast amounts of data associated with the electronic mock-up of large aerospace components, within the ever-decreasing development schedules required, would not be possible without some form of Engineering Data Management system.

Appendix A

Contact listing

There are many sources of help for companies interested in the feasibility of EDM technology or implementing a specific EDM solution. Rather than compiling a long list, the accuracy of which will degrade with time, here we restrict the contacts and addresses to those major organizations with a proven track record in establishing, promoting and supplying EDM-related awareness and technology. Although the list is not split into a number of distinct sections, it should be noted that each of the entries has a main 'category' associated with it. The purpose of this is to provide an indication as to the main operational area of the contact concerned. Note that the categories may not be mutually exclusive. For example, many EDM software vendors may also provide consultancy services, while many major system suppliers may also provide systems integration services and perhaps individual software solutions. The categories are therefore provided for overall guidance only.

- *Consulting* Many consultants claim to have expertise in EDM technology, but it is difficult to assess the degree to which these claims can be met without contacting each one of them individually. Contacts here are therefore restricted to those known to have extensive, practical experience on EDM.
- *Major platform suppliers* Contacts in this category include those major system (hardware) suppliers of the past who have moved into the solutions and services aspects and who generally have written, or adopted via third party alliances, EDM software and solutions.
- *EDM software vendors* This category details those current suppliers and products which are established in the field of EDM. Although space precludes a detailed description of each market offering, a brief outline has been included. It is not intended that contacts in this category are definitive in nature, it is merely a register of those systems known to be established at the time of publication. The exclusion or lack of comment about a product should not be interpreted as demeaning its capabilities or supplier in any way.
- *Systems integrators* As patterns of computing technology change, so do

the individual players and suppliers involved. Many projects now encompass a wide range of hardware, communications, networking, software and database applications. Managing these types of projects is difficult and time-consuming—a task either considered not cost effective or beyond the capabilities of the majority of business enterprises. Systems integration (SI) specialists have grown to fill this gap. They are usually large computer companies, many of whom have grown from a pure hardware background, who possess the skills, resources and experience in integrating the business and computing requirements of many companies, ranging from the small/medium sized to the very large.

- *Standards bodies* This category includes those organizations which can be contacted to provide more information on their related area of expertise. Major standards organizations are mentioned, but those of a more specialist nature are also included.
- *General* This category provides contacts who may prove useful in sourcing information which is related to EDM concepts and technology.

Company name	Adra Systems—MATRIX Division
Address	Adra Systems Inc.
	59 Technology Drive
	Lowell
	MA 01851
	USA
Telephone	508–937–3700
Fax	508–453–2462
Contact	Marc Warren, Groupware Marketing Manager
Category	EDM Software Vendor
Application/Product Name	MATRIX

Brief details The MATRIX division of Adra was formed in 1992 and is dedicated to the development of PDM products and solutions. The system is positioned as an 'advanced information management suite' and is designed to guide users to sets of information using highly visual navigation techniques. This visual user interface provides browsing tools and viewers which make the exploration and retrieval of information quick and easy. A powerful set of control and management structures exist below this user interface to allow vast amounts of documentation and ongoing changes to be administered. The MATRIX system can control the most complex relationships between objects regardless of source, file type or computer application. Its particular strengths are quoted as being: defining and controlling data objects; modelling organizations, groups and roles: visualizing, managing and changing processes; integrating existing, new or planned system facilities.

Company name	**Andersen Consulting**
Address	2 Arundel Street
	London
	WC2R 3LT
	United Kingdom
Telephone	0171-438-5000
Fax	0171-831-1133
Contact	Ms Betty Thayer, Associate Partner
Category	Consulting
Application/services	Industrial Consulting

Brief details   Andersen Consulting is one of the largest consulting firms in the world, with over 2000 consultants specializing in industrial projects. It has offices in more than 25 countries, and can provide worldwide support for a wide range of complex business change projects.

Company name	**ANSI**
Address	1430 Broadway
	New York
	NY 10018
	USA
Telephone	(212) 642-4900
Fax	
Contact	
Category	Standards Body
Brief details	See Glossary.

Company name	**British Computer Society**
Address	1 Sanford Street
	Swindon
	Wiltshire
	SN1 1HJ
	United Kingdom
Telephone	01793–417417
Fax	
Contact	
Category	Standards Body

Company name	**British Standards Institution**
Address	IT Services
	Linford Wood
	Milton Keynes
	MK14 6LE
	United Kingdom

Telephone	01908 220022
Fax	
Contact	Enquiry Desk—Extension 2751
Category	Standards Body
Company name	**Bull S.A.**
Address	7 rue Ampere
	91343 Massy Cedex
	France
Telephone	33 (1) 6993–8061
Fax	33 (1) 6993–8039
Contact	PDM International Competence Centre
Category	System Integrator
Application/product name	The Bull PDM Solution

Brief details Bull's PDM Solution provides the necessary framework to implement Product Data Management technology to improve a company's competitive advantage. Bull's background in manufacturing, their systems integration skills, and their use of a leading-edge PDM system, combine to make a solid solution for the PDM marketplace. The solution provides features such as process workflow management, permissions and authorizations for data access, configuration management, data exchange, classification and information viewing. It also provides all necessary tools to develop application interfaces in order to leverage current technology investments. The solution targets organizations of all sizes in multiple industry sectors, such as aerospace, defence, automotive, electronics, pharmaceuticals, food, petrochemicals, energy, transportation, healthcare.

Company name	**CGSA**
Address	8 Canalside
	Lowesmoor Wharf
	Worcester
	WR1 2RR
	United Kingdom
Telephone	01905–613236
Fax	01905–29138
Contact	Membership Secretary
Category	General

Brief details The Computer Graphics Suppliers' Association is a professional association representing the UK CAD/CAM, computer graphics and document management industries. Since being established in 1985 the CGSA has grown rapidly and now boasts some ninety member companies representing all aspects of the industry. The association provides a wide number of activities and services including: legislative and corporate issues, confidential information exchange, sponsorship of promotional events, market education, legal and industry advice, special interest groups, standards information and industry awareness and training.

Company name	CIMdata Inc.
Address	3893 Research Park Drive
	Ann Arbor
	MI 48108
	USA
Telephone	(313) 688-9922
Fax	(313) 688-1957
Contact	Ed Miller, President
Category	Consulting
Application/services	PDM Marketing and Consultancy

Brief details CIMdata specialize in the application of computers to engineering and manufacturing industry, and provide in-depth market research and technical consulting for system suppliers and end users on a worldwide basis. Although particular expertise has been gained in PDM (Product Data Management), the company's expertise spans a wide range of manufacturing systems and technologies including CAD/CAM/CIM, solid modelling, numerical control software and manufacturing resource planning (MRP). CIMdata publish a wide range of reports and technology guides including specialist PDM-related publications, for example a PDM Buyer's Guide, Technology Guide, pricing analysis and conference proceedings. CIMdata have offices in the USA and Europe and have clients in North America, Europe and the Pacific Rim.

Company name	CIMLINC Ltd.
Address	Highfields Science Park
	University Boulevard
	Nottingham
	NG7 2RQ
	United Kingdom
Telephone	0115–925–6255
Fax	0115–925–2620
Contact	Geoff Knight, Product Manager
Category	EDM Software Vendor
Application/product name	LINKAGE

Brief details CIMLINC's LINKAGE is a software tool which allows users to create applications which gather data automatically from different networked computer systems and combine it into user-defined screen formats. As such it is more of a development toolkit/environment, rather than a typical EDM 'system'. Application development is done in an open client/server environment. All information and human expertise relating to an engineering activity is accessed live, and merged together for immediate use. Graphical process planning, manufacturing instructions and information retrieval systems for engineers are typical applications developed within LINKAGE. LINKAGE integrates different computer applications, such as CAD/CAM, MRP, EDM and image processing. Information produced in LINKAGE can contain text, vector/raster graphics, audio and keyboard input data.

Company name	**Cimtech Ltd**
Address	University of Hertfordshire
	45 Grosvenor Road
	St Albans
	Hertfordshire
	United Kingdom
Telephone	01707–284691
Fax	01707–284679
Contact	Tony Hendley, Technical Director
Category	General/consulting

Brief details Cimtech provides impartial information and advice on all aspects relating to the control and management of records and documents. It is a wholly owned subsidiary of the University of Hertfordshire and has been established for over 20 years. Cimtech is highly regarded worldwide for its informed and unbiased assessment of records management, policies and procedures, file structures, classification and indexing systems, storage media, hardware and software. Clients include BAe Dynamics, CCTA, Rediffusion Simulation, Nuclear Electric and the UK Government Department of Trade and Industry.

Cimtech's expertise is made available to both members and non-members via a wide number of publications, and their educational, enquiry, and comprehensive range of consultancy services.

Company name	**Computervision Ltd**
Address	Argent Court
	Sir William Lyons Road
	Coventry
	CV4 7EZ
	United Kingdom
Telephone	01203–417718
Fax	01203–692 418
Contact	Sarah Farron, Sales Program Coordinator
Category	EDM Software Vendor
Application/product name	EDM

Brief details Computervision have been very successful in the EDM market-place, although this has occurred primarily within their established CAD/CAM customer base and are not as generally visible as some other vendors. Their EDM product is a modular solution that combines RDBMS technology with task-orientated applications. The four main modules are EDMVault, EDMClient, EDMProjects and EDMProgramming, and have evolved from initial 'toolkits' into sophisticated products in their own right. The user interface is comprehensive and easy to use, and the integration to the associated graphics modules is seamless and natural, although the product can also be used in a more generic sense. The EDM solution provides support for heterogeneous system environments and CSAs and is widely regarded as being a solution capable of providing enterprise-wide implementation capability.

Company name	**DTI**
Address	Management & Technology Services Division
	151 Buckingham Palace Road
	London
	SW1 9SS
	United Kingdom
Telephone	0171–215–5000
Fax	
Contact	DTI Enquiry Desk
Category	General

Company name	**ECS Ltd**
Address	Aldershawe Hall
	Claypit Lane
	Wall
	Nr Lichfield
	Staffordshire
	WS14 0AQ
	United Kingdom
Telephone	01543–414751
Fax	01543–250159
Contact	Nicola Baxter, Marketing Manager
Category	EDM Software Vendor
Application/product name	Document Manager

Brief details ECS have extensive experience in the CAD/CAM marketplace and claim to have the third largest installed base of any CAD/CAM supplier in the UK. They have a marketing agreement with Cimage International Ltd to market the company's Document Manager product in the UK. ECS also have experience in connectivity products and general engineering consultancy services.

Company name	**Eigner & Partner**
Address	Ruschgraben 133
	76139 Karlsruhe
	Germany
Telephone	++ 49–721–6291–78
Fax	++ 49–721–6291–88
Contact	Werner Teubner, Partner Distribution
	Department
Category	EDM Software Vendor
Application/product name	CADIM

Brief details   Eigner & Partner are a well-known EDM supplier in Europe (especially in their home market) and are now expanding their operations into the USA and Asia. Their product is called CADIM (Computer Aided Data Integration Manager) and is a modular, hardware and database-independent product data management system. The product provides extensive functionality including the ability to manage documents, BOMs, and part master data, together with workflow management and advanced viewing capabilities. Optional functionality includes variant BOMs, classification and coding systems (group technology), decision tables, workplan management and archiving systems. The system is available in two upward-compatible forms (Startup and EDB). This facilitates its introduction into companies wishing to start with drawing file management and move onto a full EDM solution at some later stage. CADIM is available on a wide range of different hardware and database options (e.g. Ingres, Informix, Oracle, RdB and Sybase) and also has a wide number of system interfaces available as standard.

Company name	**Hewlett-Packard Company**
Address	3404 East Harmony Road
	Ft Collins
	Colorado 80525
	USA
Telephone	(303) 229–3800
Fax	(303) 229–7182
Contact	Debbie Madden, Marketing
	Communications Manager
Category	EDM Software Vendor
Application/product name	HP WorkManager

Brief details   HP have been involved in EDM for a number of years with their HP-DMS product, which was mainly focused on the engineering department support of ME-10 installations. HP WorkManager was introduced in 1992 and extends this established presence into enterprise-wide product data and workflow management. The system coordinates the entire product design and manufacturing lifecycle, from the initial BOM and drawings to the review and approval process, prototype management and the integration of applications. It integrates closely with Hewlett-Packard's SolidDesigner CAD system and uses user-defined methods to manage the content, flow and processing of product development data. HP WorkManager operates in a distributed environment and facilitates the sharing of data to enable a true concurrent engineering environment to exist.

Company name	**IBM**
Address	Consulting Group
	Production and Engineering Business
	PO Box 31
	Birmingham Road
	Warwick
	Warwickshire
	CV34 5JL
	United Kingdom
Telephone	01926–464000
Fax	01926–407215
Contact	Ray Channell, Engineering Consultant—
	Production and Engineering Business
Category	EDM Software Vendor
Application/product name	ProductManager

Brief details   Some consider IBM to have had a confused and complex range of engineering data management products in the past. However, this has now changed with the introduction, in mid-1992, of ProductManager, their premier application family for product data management. ProductManager consists of three main licensed program products—Application Services Manager, Product Change Manager and Product Structure Manager—which together provide a comprehensive data management solution which has been well received by the marketplace. Versions are available on the MVS mainframe operating environment as well as on the RS/6000 running AIX and both are capable of supporting large-scale implementations. The 6000-based system is available with an OSF/Motif graphical user interface and both can take advantage of an optional product for viewing and redlining called EXPRESSight/6000.

Company name	**Institute of Configuration Management**
Address	PO Box 5656
	Scottsdale
	AZ 85261–5656
	USA

Telephone	++ 602–998–8600
Fax	++ 602–998–8923
Contact	Vince Guess, President
Category	General

Brief details   The Institute of Configuration Management (ICM) evolved from an effort initiated in 1978 to apply continuous improvement to the configuration and change management process. A proven set of techniques evolved and enhancements have been continuous. ICM provide a wide range of consultancy and Configuration Management II (CMII) awareness services. Their training courses are modular in structure and can lead to a full certification in the CMII process.

Courses are also administered in Europe through European Configuration Management II Ltd—contact Callum Kidd on 01532–332634.

Company name	**Metaphase Technology, Inc.**
Address	4201 Lexington Avenue North
	Arden Hills
	MN 55126
	USA
Telephone	++ 612–482–2170
Fax	++ 612–482–4348
Contact	Joseph S. Malloni, Manager,
	Product Marketing
Category	EDM Software Vendor
Application/product name	Metaphase 1.0

Brief details   Metaphase 1.0 is a full suite of well-packaged products and technologies designed to address enterprise-wide Product Data Management. It is provided for sale 'off-the-shelf' through resellers or as a flexible technology by application vendors. The Metaphase product provides:

- Network object management across a distributed environment (either peer-to-peer or hierarchical)
- Process definition for the sharing and distribution of data objects throughout a network of groups of users
- Product structure and configuration management
- Image services for CALS compliant viewing and markup functionality
- Output services for printing and plotting
- Integration toolkit for encapsulations and integration with the Metaphase PDM framework. This includes both a command line and 'C' callable interfaces

The Metaphase product is available on a wide range of hardware platforms and is considered to be one of the market's leading players.

Company name	**MoD CALS**
Address	UK CALS News Office
	Directorate of Procurement Policy (Studies)
	Room 6302
	MoD Main Building
	Whitehall
	London
	SW1A 2HP
	United Kingdom
Telephone	0171–218–7515
Fax	
Contact	Mr Robert Surtees, Studies Team Member
Category	General

Company name	**MSPL Ltd**
Address	Beaumont
	Burfield Road
	Old Windsor
	Berkshire
	SL4 2JP
	United Kingdom
Telephone	01753–833844
Fax	01753–833026
Contact	Peter Cefai, Sales Manager
Category	EDM Software Vendor
Application/product name	EMCS-X

Brief details   EMCS-X is an EDM system designed for use by engineering departments and for onward transmission to the production function, typically an MRP system. It is a UNIX-based application written in Ingres 4GL. EMCS-X is a modular system, the entire solution currently comprising eleven individual elements which can integrate together to provide a comprehensive solution. The modules are: parts registry, BOM, engineering change control, work routes, file management, mail messaging, build standard control, audit trail, security and administration, and internetworking which consists of two interface modules—one to CAD and the other to MRP.

Company name	**OSF**
Address	11 Cambridge Center
	Cambridge
	MA 02142
	USA

Telephone	(617) 621–8700
Fax	
Contact	
Category	Standards Body
Company name	**PA Consulting Group**
Address	Cambridge Laboratory
	Melbourn
	Royston
	Hertfordshire
	SG8 6DP
	United Kingdom
Telephone	01763–261222
Fax	01763–260023
Contact	Mr Barry Brooks, Managing Consultant—
	Product and Process Engineering
Category	Consulting
Application/services	Engineering and Technology Consulting

Brief details   PA Consulting Group is a leading international management and technology consultancy. PA work with clients in industry, commerce and government to manage complex change and to create business advantage through: enhanced strategic thinking, achieving sustained customer satisfaction, utilizing the power of appropriate technology, using information effectively, realizing the potential of people, and raising overall performance.

PA's Global Technology Group in particular provides a wide range of professional services associated with the design and development of products and processes. These include concurrent engineering, advanced manufacturing technology, Engineering Data Management, computer graphics applications, and technology management. PA is located in 66 offices and 20 countries and employs 2300 staff of the highest calibre.

Company name	**PAFEC Ltd**
Address	Strelley Hall
	Nottingham
	NG8 6PE
	United Kingdom
Telephone	0115–935–7055
Fax	0115–935–7057
Contact	Martin Bayton, Marketing Manager
Category	EDM Software Vendor
Application/product name	PAFEC EDM

Brief details   PAFEC have an established history in the provision of software solutions for engineering analysis and computer-aided engineering. Their

EDM solution builds on this success and is designed to meet industry's growing need for information control. It is capable of managing a diverse range of engineering data and controlling engineering processes such as change control, release control, product configuration and archiving. It is a modular system which provides a flexible structure to allow customized solutions to be easily built. Document scanning, printing and plotting software options are also available.

Company name	**SDRC**
Address	Milford House
	Priory End
	London Road
	Hitchin
	Herts SG4 9AL
	United Kingdom
Telephone	01462–440222
Fax	01462–440522
Contact	Stephen Grey-Wilson, PDM Consultant
Category	EDM Software Vendor
Application/product name	Desktop PDM

Brief details   Desktop PDM is a modular software package that assists manufacturers in managing work-in-progress product information. It combines framework and PDM (Product Data Management) capabilities that allow users to access, search and retrieve product information as it is generated throughout the design and manufacturing process. SDRC is a leading international supplier of mechanical design automation and PDM software products, as well as consulting services. They offer enterprise-capable EDM system solutions and have focused attention on providing support for heterogeneous system environments, distributed CSAs and support for large-scale implementations.

Company name	**ServiceTec Infographics Ltd**
Address	11 Dunlop Square
	Deans Industrial Estate
	Livingston
	West Lothian
	EH54 8SB
	United Kingdom
Telephone	01506–411583
Fax	01506–412336
Contact	David Mercer, Manager—
	International Sales
Category	EDM Software Vendor
Application/product name	MAZURKA Data Manager

Brief details   ServiceTec International Plc is in the business of supplying, installing, commissioning, project managing and supporting integrated solutions which span the entire spectrum of the modern information technology world. The Data Manager product provides engineering data control and administration functions for a complete CAD/CAM environment. It is a modular system which provides product structure definition and interrogation, document filing, change management, configuration control and project monitoring for task identification and cost monitoring purposes.

Company name	**Sherpa Corporation**
Address	Doncastle House
	Doncastle Road
	Bracknell
	Berkshire
	RG12 8PE
	United Kingdom
Telephone	01344–867222
Fax	01344–868199
Contact	Tony Appleby, Director of Sales
Category	EDM Software Vendor
Application/product name	DMS/PIMS

Brief details   Sherpa Corporation was incorporated in 1984 and defined its long-term mission as the development of product information management (PIM) solutions. The company's initial products were centred around EDM development tools and procedures—these have now developed into a mature set of product information management systems and solutions which span the entire product life-cycle. Sherpa are an established supplier in the EDM marketplace, capable of providing enterprise-wide solutions, and claim to have the largest 'independent' installed PIM user base. Sherpa's product is considered to be a good 'all-rounder', providing a balanced solution over most commonly accepted areas of functionality. They have considerable experience in heterogeneous environments and CSAs, and have invested in comprehensive sales, technology awareness and training initiatives.

Company name	**UKCIC**
	(United Kingdom CALS Industry Council)
Address	c/o EEA
	Russell Square House
	10–12 Russell Square
	London
	WC1B 5AE
	United Kingdom

Telephone	0171–331–2017
Fax	0171–331–2040
Contact	Mr David Froome, Chairman
Category	General

Brief details The UKCIC is an association of major UK trade associations and other bodies, dedicated to the maintenance and improvement of UK industry's competitive position in world markets, through the adoption of CALS or CALS-like strategies. UKCIC is active internationally in both the USA and Europe, representing the need for international standards and bringing back details of future trends and current best practice.

Company name	**Workgroup Technology Corporation**
Address	81 Hartwell Avenue
	Lexington
	MA 02173
	USA
Telephone	++ 617–674–2000
Fax	++ 617–674–0034
Contact	Theresa Delfino, Manager,
	Marketing Programs
Category	EDM Software Vendor
Application/product name	CMS

Brief details CMS is a Product Data Management and Workflow solution that manages all product-related data, from virtually any application or source, throughout the entire product life-cycle. The product runs on a wide variety of hardware platforms and links to Oracle and Sybase RDBMS. It provides an exceptionally good user interface via an intuitive graphical user environment.

CMS/Workflow, an add-on module to CMS, is a dynamic tool for re-engineering business processes and procedures. CMS/Pro provides a direct interface to Parametric Technology's Pro/ENGINEER, allowing design teams to share design information and collaborate on projects.

CMS is used for mission-critical applications, such as engineering change control, regulatory compliance (ISO 9000 or FDA), product modelling, and on-line file access for remote locations.

Appendix B

Glossary

The primary aim of this glossary is to act as source of further information on many of the terms used within the main body of text, even although the terms are often briefly explained as they are introduced. The descriptions given should not be considered as formal definitions—they are included as explanations only.

Any technology such as EDM suffers from the use of many acronyms. Their use has been limited as far as possible, but to a large extent they have become an inevitable fact of life. Acronyms (or TLAs as they have been called thus far!) can prove to be a problem for those who do not know what they mean, and so have been given priority over the full explanation of the term.

As was mentioned in the Preface, this book was intended to provide a broad introduction in a practical, and as far as possible non-technical, manner to allow the reader to explore the various aspects of the subject in more detail. It is hoped that the glossary assists in introducing this 'second level' of detail.

The volume and content of the glossary has been targeted specifically to that which has been discussed in the body of the text and kept to a reasonable level. There are many other entries which could have been included—indeed there are many other glossaries (for example of IBM system technology) which are longer than this entire book. It is hoped that the few additional items which have been included are relevant and that the glossary itself proves useful.

AECMA Association Europeénne des Constructeurs de Matériel Aerospatial. An international standards body whose range of specifications can be used as standards for the communication of technical information within the CALS strategy.

Alphanumeric display/terminal A character-based terminal used to display computer-generated information (formerly known as a visual display unit).

ANSI The American National Standards Institution. ANSI is a privately funded, non-profit making organization that administers the US private sector voluntary consensus standards system.

API Application programming interface. An API is a standard interface between an external application program and a set of services, such as those provided by the functions and database capabilities of an EDM system. The interface is usually built from a library of routines that can be embedded within the code of the application itself.

Applications program A program or set of programs designed for some specific application.

Approval A process whereby changes to individual documents, groups of documents, or controlled data are approved by the appropriate person(s).

Architecture This is a broad term that denotes a description of some significant part of a computer system or of a particular aspect of computing, e.g. development or operations.

Archival storage The storage of computer-generated data in a more or less permanent form, e.g. magnetic tape. Unlike backup storage, such data is not normally available on-line.

ASCII American Standard Code for Information Interchange.

Authorization Within an EDM environment, authorization usually refers to the level of functionality and access given to specific classes of system user.

Backup storage Information stored on backup devices such as disks which can be accessed very quickly (i.e. 'on-line'). Also known as virtual memory.

BOM Bill of materials. An ordered list of parts, assemblies and raw materials that define the content of a particular product. Also known as a product structure, it defines the number, type, quantity, and relationships between the various levels.

BSI British Standards Institution. The UK national standards setting body.

C A programming language with exceptional portability. It is used extensively to program software that is intended to be ported to many different platforms.

C++ A variant of C that is used for programming object-orientated systems.

CAD/CAM Computer-aided design/computer-aided manufacturing. CAD is sometimes also known as computer-aided draughting. All are important applications in the field of engineering computers, are graphically-orientated, and run on high-end PCs and UNIX-based workstations.

CAE Computer-aided engineering. A general term used to describe the application of computer-based solutions to engineering tasks, e.g. CAD, CAPP, FEA.

CALS Continuous Acquisition and Life-cycle Support (previously known as Computer Aided Acquisition and Logistics Support). CALS is a US DoD Department of Defense program to acquire and manage technical information in electronic form. It is now spreading into commercial environments and includes standards for electronic data interchange, electronic technical documentation and guidelines for process improvement.

CAPE Concurrent art to product environment. A model established by the

Gartner Group in 1991 that describes the essential technology elements and user requirements necessary to implement a world-class, enterprise-wide product development environment.

CAPP Computer-aided process planning.

CCITT The Comité Consultatif Internationale de Télégraphique et Téléphonique. The CCITT is part of the International Telecommunications Union, an agency of the United Nations.

CD-ROM Compact Disc-Read Only Memory. A storage medium typically used for large amounts of information.

CE Concurrent engineering. A management methodology in which personnel from all disciplines (e.g. marketing, design, production, etc.) work together through all phases of the product life-cycle to improve quality, competitiveness and responsiveness to market requirements. Also known as simultaneous engineering.

CFD Computational fluid dynamics. The study of fluid flow and heat transfer problems in a computational domain. General CFD problems may be solved using finite element analysis (FEA), finite difference, or finite volume methodologies.

Check-in/out The process of retrieving or returning EDM-controlled data objects and documents to a secure storage area often known as an electronic vault.

CIM Computer-integrated manufacture.

Client/server A method of distributed processing where client systems process information locally using their own computing power. Servers hold database and other information which can be accessed by multiple clients.

CM/CMII Configuration management. The process of defining and controlling a product structure and its related documentation with particular emphasis on revision control and the maintenance of associated historical information.

CNC Computer numerical control. A method of controlling a machine tool which utilizes a computer to calculate the required cutter paths and other appropriate information.

Compiler A program which translates a high-level language, such as Fortran or C, into machine code.

DBMS Database management system. A computer tool to store information in an independent and logical manner.

Decnet Digital Equipment Corporation's network architecture.

De facto (standards) *De facto* standards are those that have arisen through general wide usage rather than through any conscious standards-setting process.

Disk drive A device which allows information to be written-to and read from magnetic disks at high speeds.

DRM   Design release management. DRM is a generic term to describe the process of controlling design data within an EDM environment. It includes check-in/out, access control and authorization processes.

DTI   The Department of Trade and Industry. A UK government department (equivalent to the US Department of Commerce) which has adopted a policy of promoting open systems.

ECN   Engineering Change Note. A formal document indicating the acceptance of a previously defined change.

ECR   Engineering Change Request. ECRs are documents which notify selected persons of proposed or pending changes.

EDI   Electronic document interchange. The exchange of commercial documents (e.g. orders, invoices) by electronic means.

EDM   Engineering Data Management.

EDMICS   Engineering Data Management Information and Control System. The term for EDM used within the CALS/ILS (Integrated Logistics Support) environment.

Effectivity   These are indicators in a product structure which notify, by either the dates, serial numbers or batches/lots, which parent/child conditions apply at any specific time.

Ethernet   A type of LAN structure originally developed in 1981 by Xerox, DEC and Intel. The associated protocols have been subjected to standardization as IEEE 802.3.

FEM/FEA   Finite element meshing/finite element analysis. The finite element technique involves dividing a large or complex object into a number of smaller separate elements and then considering any forces acting on each element in turn. In this way it is possible to calculate where and when the object may fail under physical or thermal stress.

FMS   Flexible manufacturing system. An integrated production unit usually consisting of a few machine tools, transfer mechanisms and robots, all under computer control.

Group technology   A methodology for the optimal production of like items, based on their initial classification and coding.

GUI   Graphical user interface. A user interface style in which graphics are used as the primary means of communication.

Host computer   A central computer which controls a number of remote terminals or workstations.

Icon   A graphical image on a computer screen used to denote an object that the user can interact with.

IEE   The Institution of Electrical Engineers. The UK professional body for manufacturing and electrical engineers.

IEEE   The Institute of Electrical and Electronics Engineers. The US equivalent of the IEE in the UK. It is a professional organization that sets standards for computers and communications.

IGES   Initial Graphics Exchange Specification. IGES is an encoding format for the exchange of engineering drawings and associated information in digital form.

Interoperability   The ability of computers from different vendors to work together.

IRR   Internal rate of return. An approximate measure of the rate of return of a project or contract as used in investment appraisal. It is a discounted rate that equates to the present value of the revenues, with the present value of the costs.

ISO   The International Standards Organization. An international group that sponsors the development and acceptance of voluntary standards.

IT   Information technology. IT is a term used predominantly in the UK to describe computer-based technology. It is broadly equivalent to IS (Information Services) and MIS (Management Information Systems) as used in other countries.

LAN   Local Area Network. A generic term for a localized network of small computers (such as PCs) and associated peripherals (typically connected using Ethernet, token ring, or other protocols) through a central control unit, allowing users in an office, building, or locality, to share programs, information and equipment.

MAP   Manufacturing Automation Protocol. A set of standards developed by General Motors for use in production automation.

Metadata   Information about the data objects controlled by an EDM system, for example part number, quantity, description. Such data is usually defined and controlled from within the DBMS (the core element of an EDM system) via a database model or 'schema'.

MIL-SPEC   US DoD Military Specification.

MIL-STD   US DoD Military Standard.

MIS   Management Information System.

MRP/MRPII   Material Requirements Planning/Manufacturing Resource Planning. Powerful manufacturing tools which give management the control required over a business to plan for future requirements.

MS-DOS   A PC operating system developed by Microsoft for 16 bit machines that has become a *de facto* industry standard.

MVS   An IBM operating system for 370-architecture (large) machines.

NCC   The National Computing Centre. A UK body set up with government support to act as a centre of computing expertise.

NIST   National Institute of Science and Technology. A US body formerly known as the National Bureau of Standards. NIST is a federal agency that cooperates with IEEE and also sponsors OSI workshops.

NPV   Net present value. A measure of investment appraisal usually calculated as the sum of the discounted cash flows, both positive and negative, over the life of a project or contract.

Object-orientated system   A system or database that deals primarily with objects ('encapsulated' items of data), rather than simply raw data, images, text, etc., explicitly.

OLTP   On-line transaction processing. A generic term for applications which involve a number of terminals entering a large number of transactions that have to be processed in a very fast and efficient manner.

Open systems   Systems that do not conform to an architecture that is significantly controlled by one single vendor.

Operating system   The program which administers the computer's own internal workings and peripherals, and allows it to understand the applications programs that are loaded into it.

OSF   The Open Software Foundation. An alliance of system suppliers to create standards for the UNIX operating system. OSF is an international non-profit making organization formed in 1988 to develop an 'open' software environment, to ensure interoperability of different vendors' hardware and software. It is based in Cambridge, MA.

OSI   Open Systems Interconnection. A suite of standards to specify a system-independent method of communication. OSI is a reference model or logical structure around which an open system architecture may be built. The ISO 7498 reference model specifies a seven-layer network architecture which enables any OSI-compatible device to communicate fully with any other OSI compatible device.

Payback period   A calculation, based on the operating contribution before interest expenses, of the time period required for the cumulative cash flow to turn positive, and remain positive for the duration of the contract/project.

PDES   Product Data Exchange using STEP (originally Product Data Exchange Specification). The US version of STEP. Also, PDES Inc. is a non-profit making organization aimed at accelerating the commercial implementation of STEP.

Platform   A general term for a range of hardware and/or systems software.

Portability   The ability of an application program to run on computer systems from many different vendors and with different architectures.

Protocol   A convention defined to allow the correct interpretation of information.

RAM   Random access memory. The main element of computer memory (used to be known as core memory) in which any address can be accessed with equal ease and speed.

RDBMS A relational database management system. A DBMS which uses table indices to maintain data records and associated relationships.

RISC Reduced Instruction Set Computer. An approach to designing computer instruction sets to produce cheaper machines with more power. Until fairly recently most computers were designed as CISC (Complex Instruction Set Computer) machines.

Scalability The capability of using the same operating system or software on many classes of computer, from PCs to mainframes.

SGML Standard Generalized Markup Language. A set of syntax and semantics used to describe the allowed structure of a class of documents.

SQL Structured Query Language. A database query language originally developed by IBM and now subject to international standardization.

STEP The Standard for the Exchange of Product Model Data. An international standard, originally developed by NIST and now adopted by the ISO, as a means of transferring product data between dissimilar systems in a standard format. Although it is only currently in draft form, it promises to be superior to other data exchange standards such as IGES and DXF (originally developed by Autodesk).

TCP/IP Transmission Control Protocol/Internet Protocol. The two main *de facto* network communication protocols—part of the Internet Protocol Suite developed under the auspices of the US DoD. These protocols have gained wide usage, particularly in UNIX environments.

Terminal A device used to display input and output from the computer system it is connected to.

TOP Technical and Office Protocol. A set of standards for office systems originally defined by the Boeing company.

Trigger A mechanism which detects activity or change related to a data object and can initiate some subsequent (but predefined) action.

UNIX A computer operating system originally developed by Bell Telephone Laboratories (AT&T) which has been adopted by the open systems movement as the base operating environment for a wide range of machine sizes and configurations. UNIX enables many types of computers to share programs, data and peripherals.

Vault A secure area for the central control and filing of EDM-related data which is subject to specific rules and processes. Also known as an electronic vault.

Version A numeric or alphanumeric code which is used to represent structures and objects as they change during their life-cycle. Also known as part/structure revisions.

WAN Wide Area Network. To be distinguished from a LAN, a WAN interconnects sites which are geographically remote. WANs typically (although not necessarily) use much lower transmission speeds than LANs.

WBS Work breakdown structure. A mechanism for breaking down work into smaller, more manageable elements. Often used as a basis for controlling projects.

Workgroup A team or group of people working on the same project or towards a common goal.

Workstation A graphical device connected to a server on a network allowing the user to transfer information and work off-line.

X/Open An international consortium of computer vendors charged with creating a common applications environment based on the most useful formal and *de facto* standards available.

X-Windows A graphical user interface developed at the Massachusetts Institute of Technology (MIT) which has received wide acceptance.

Bibliography

Boothroyd, G. and Dewhurst, P., *Product Design for Assembly*, 1990. Boothroyd Dewhurst, Inc., Wakefield, R.I.

Computervision, *The 1993 Manufacturing Attitudes Survey—Management Summary*, 1993. Computervision Ltd, Coventry.

Data General Ltd, *Practical Open Systems—Downsizing: The Business Opportunities*, 1992. KPMG Management Consulting, London.

Date, C. J., *An Introduction to Database Systems*, Volume I, Fourth Edition, 1986. Addison-Wesley, Reading, MA.

Digital Equipment Corporation, *Introduction to Local Area Networks*, 1982. Digital Equipment Corporation.

DTI, The Enterprise Initiative, *Managing into the '90's—Manufacturing*, 1990. DTI, London.

Frost & Sullivan Inc., *Engineering Data Management—Implementation Strategies in Europe*, Winter 1989–1990.

Harrison, F. L., *Advanced project management. A structured approach*, 1992. Gower, Aldershot.

Hugo, Ian, *Practical Open Systems—A Guide For Managers*, 1991. Data General Limited/NCC/Blackwell, Oxford.

Kharbanda, O. P. and Stallworthy, E. A., *Project Teams. The Human Factor*, 1990. NCC Blackwell, Oxford.

Martin, James, *An End-User's Guide To Data Base*, 1981. Prentice-Hall, Englewood Cliffs, NJ.

Medland, A. J. and Burnett, P., *CADCAM in Practice*, 1986. Kogan Page, London.

PA Consulting Group, *CADCAM—opening the door to CIM*, 1988. PA Consulting Services Ltd, Royston, Herts.

Primrose, P. L., *Investment in Manufacturing Technology*, 1991. Chapman and Hall, London.

Reiss, Geoff, *Project Management Demystified*, 1991. Spon, London.

Rodgers, Ulka, *Unix Database Management Systems*, 1990. Prentice-Hall, Englewood Cliffs, NJ.

Schofield, N. and Bishop, T., *Unlocking the Potential of CIM—A management guide*, 1989. PA Consulting Services Ltd, Royston, Herts.

Smith, Joan M., *An Introduction to CALS: The Strategy and the Standards*, 1990. Technology Appraisals Limited.

Wiederhold, Gio, *Database Design*, 1981. McGraw-Hill, Maidenhead.

Index